Quick Changeover Concepts Applied

Dramatically Reduce Set-Up Time and Increase Production Flexibility with SMED

Karsten Herr

Quick Changeover Concepts Applied

Dramatically Reduce Set-Up Time and Increase Production Flexibility with SMED

CRC Press is an imprint of the
Taylor & Francis Group, an **informa** business

CRC Press
Taylor & Francis Group
6000 Broken Sound Parkway NW, Suite 300
Boca Raton, FL 33487-2742

Printed on acid-free paper
Version Date: 20131023

International Standard Book Number-13: 978-1-4665-9631-3 (Paperback)

Library of Congress Cataloging-in-Publication Data

Herr, Karsten.
Quick changeover concepts applied : dramatically reduce set-up time and increase production flexibility with SMED / Karsten Herr.
pages cm
Includes bibliographical references and index.
ISBN 978-1-4665-9631-3 (pbk.)
1. Production management. 2. Lean manufacturing. 3. Industrial equipment. I. Title.

TS155.H3965 2013
658.5--dc23 2013039516

Visit the Taylor & Francis Web site at
http://www.taylorandfrancis.com

and the CRC Press Web site at
http://www.crcpress.com

About the author

Karsten Herr has studied brewing technology and mechanical engineering. He has worked for many years as an operations and production manager and as a commissioning engineer. Since 2004, he has been working as a freelance consulting engineer, training and coaching production companies. He supports production companies of all sectors, in discrete as well as in continuous industries. Projects on which he has worked range from complete production restructuring via optimizing individual value streams to one-off training courses on different topics on process and workplace organization.

Herr has specialized in the analysis of set-up procedures on machines and processes and the drafting of quick changeover concepts. He regularly supports companies with open and in-house workshops and analysis of set-ups and drafting of solutions for quick changeovers as well as running improvement projects of various scopes.

He also works as a lecturer at various universities and professional academies, where he teaches courses on organization and structuring of business processes, systems engineering and systems behavior, production organization and factory planning, and production optimization. His main focus is on Lean, TOC, TPM, QRM, and Six Sigma (Certified Black Belt).

Herr has a master's degree in management and organization from the TiasNimbas Business School, one of Europe's top business schools.

Preface

The advice in this book was considered and checked carefully by the author. However, no guarantees can be made. Any liability on the author's part for property damage, physical injury and asset damage is explicitly excluded.

Neither purchasing nor reading this book discharges you from the obligation to think.

This book ought to be studied, not just read!

I would like to express my thanks to all those who have helped me in the production of this book. First among these are my wife, Sylvia, and our son, Elias. Thanks for your patience during the many hours I was sitting at my desk.

I would like to thank Ir. Frits van den Elst, Elstec Engineering, Almelo/NL, Drs. Koos Held and Drs. Gerhard Buning, both of Saxion universities, Enschede/NL and Dipl.-Ing. Joachim Martin, BA Melle/GER and Dipl.-Ing. Meinrad Schmitz-Micheel, Cetec Industrial GmbH Leverkusen/GER for their valuable technical and factual comments and useful discussions.

I am also grateful to Ing. Eghard Kolste for his support with producing the layout.

Particular thanks are due to Erwin Halder KG, Otto Ganter GmbH & Co. KG, Andreas Maier GmbH & Co. KG, and PM Bearings BV for their support with the production of the text.

Introduction

Long set-up times are an obstacle to the flexible use of production resources and are therefore a core production organization issue. Thus, reducing set-up times is a major challenge for optimization leading to a Lean production system, especially when it comes to realizing the ability to serve high-mix, low-volume markets.

Quick changeover concepts are not only for large enterprises or at high costs!

Contrary to popular belief, however, set-up times are not unalterable conditions. Likewise, the quick changeover concept is neither peculiar to large industrial companies alone nor achievable only with immense financial effort. On topics such as 5S-workplace organization, Lean production, Kanban, etc., a vast offering of books, knowledge, and advice has been built up. On the topic of set-up reduction, the situation has hitherto been quite different. Implementation-oriented publications are scarce; consultants often bypass the subject or limit themselves to a few organizational steps, ending with the advice to "shorten internal steps." *How* this is to take place and with which methods you can work remain unmentioned. This book is intended to be a contribution to closing this gap.

Shortening internal steps, but how?

In the first four chapters, the book initially discusses the general dependencies of the production organization and the role of set-ups in them, introduces the concept of "systems engineering," and considers set-up and methods for shortening set-up time from a strategic perspective.

Chapters 5 and 6 explore the set-up as a process and its elements, as well as the analysis options for these, in detail.

Chapters 7 to 10 are devoted to the drafting of quick changeover concepts. After the need for communication and documentation and the financial aspects of set-up reduction campaigns are discussed in Chapters 11 and 12, Chapter 13 finally illustrates some implementation samples.

Margin with keywords for a quick overview

In the margin at the edge of each page, the main ideas are compiled as bullet points, to allow a quick overview to simplify the user's task of finding the corresponding

places in the book if he has any questions.
With the methods for analyzing set-ups and for drafting quick changeover concepts presented here, it is possible to reduce set-up times by up to 95% and thus to a minimum.
The necessary changes are frequently achievable with very low funding.
Just-in-time and just-in-sequence are therefore achievable as a means of production flexibility also for small and medium-size companies.

Competitive edge through operational excellence

Properly implemented, the quick changeover concept represents a key part of a strategy built on operation excellence and thus a sustainable achievement for a competitive edge.

1. PRODUCTION SYSTEMS IN DYNAMIC MARKETS

Today, manufacturing companies are confronted with a host of challenges. Dictated by internationalization, free trade, and not least the improved availability of information through the comprehensive distribution of Internet connections, pressure on the markets is building. Whereas until a short time ago, sellers were the leading players on the markets, the reality today is an obvious restructuring toward buyers' markets, which puts increasing pressure on companies and forces them to respond. As a consequence, conventional pricing models in which selling prices were determined by adding up production costs and the desired profit margin, are increasingly losing their credibility. In a competitive environment, the market determines the price and the performance expectation associated with it. Prices are following a falling trend, while increasing individualization of the markets is concurrently providing an increasingly complex performance expectation. Companies are expected to meet production requirements such as small-batch and single-piece runs, shortest lead times, just-in-time (JIT), and just-in-sequence (JIS) deliveries and zero defects in ever shorter product life cycles.

From seller's market to buyer's market

Individualization of the markets, more complex performance expectations

JIT and JIS

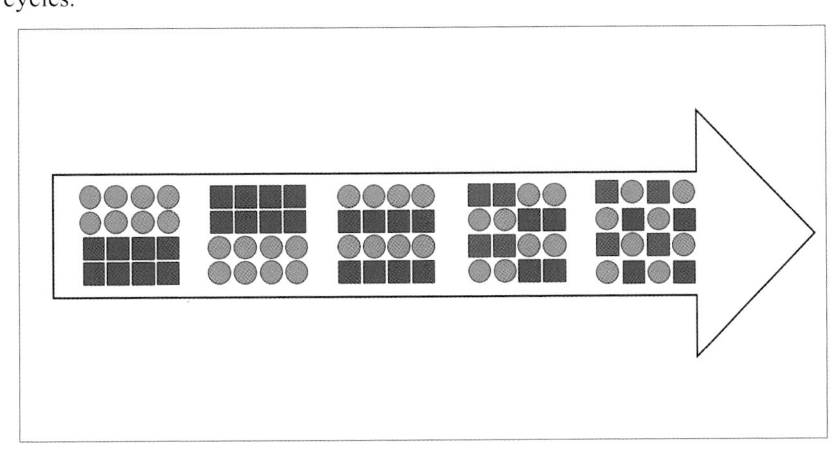

Figure 1-1: From batch & queue to JIT and JIS.

Reduce capital circulation times.

At the same time, more stringent equity rules for banks are leading to difficulty in obtaining funds.

For supply chains overall and for the companies taking part in them, this means that (internal) structures must be adapted in a fashion such that costs are lowered, the need for working capital is minimized and, concurrently, production logistical performance parameters are improved. In the past decades, this led to various new views and insights into the organization of value creation.

Improve production logistic performance parameters.

Supply chain management (SCM) is now widespread over the entire supply chain. Due to the logic of the model, the partners involved (suppliers, dealers, logistic services providers, clients) are coordinating and cooperating to produce with minimum stocks at all points of the chain, in order to speed up the overall process, become more flexible and reactive, and tie up less capital. SCM integrates management within a company's boundaries and extends it beyond them.

An important paradigm in this scenario is that individual companies are no longer competing with each other — instead, it is now networked supply chains, for which integration and coordination of the members of the "supply chain" system becomes necessary. SCM takes over this task. The level of development of SCM depends frequently on the size of the participating companies. The implementation of such systems is much more advanced in supply chains distinguished by large companies, such as in the automobile industry, in comparison to cooperating small and medium-size companies.

In individual companies and factories, concepts and models are increasingly being implemented that are based on a holistic approach and target on maximum production effectiveness under the condition of maximum efficiency. In this context, effectiveness means to do "the right things" while efficiency relates to "doing the things right." Doing something wrong properly or, in other words, working efficiently but ineffectively, is not purposeful or in the words of Russel Ackoff, "doing the

wrong thing right, means you are getting even worse."
The requirement of effectiveness under the efficiency condition can be summarized under the term "operational excellence", a term which has become increasingly used in recent years to describe an integrating term for various concepts, although there is no general definition, as far as the author of this text is aware. An Internet search will reveal a variety of definitions.
At this point, an attempt will be made to present a general definition:

Define operational excellence.

Operational excellence describes a (production-) system's adaptive ability in regard to the realization of customer value.

Customer value

The term customer value in this context has a prominent position, because it represents the external target coordinate on which internal organizational structures must be focused.

Shorten lead times.

The best known of these operational excellence concepts are lean production according to the principles of the Toyota Production System (Lean Manufacturing), the bottleneck theory (Theory of Constraints) according to Goldratt, the CONWIP system by Spearman and Hopp, and Quick Response Manufacturing according to Rajan Suri with their control systems Kanban, DBR, Conwip and POLCA.

Synchronize processes.

The focus of these concepts is in maximizing output and shortening lead times. The primary work mechanisms are active stock control in the value stream and synchronization of the individual processes that make up the value stream. It is the application of the findings of the queuing theory stating that the lead time of a system is a function of the inventory and that the development of the lead time follows varying laws depending on the degrees of utilization of the parts of the system.

Inventory control with Kanban

By introducing production control concepts based on inventory control, such as pull-production (Kanban) and the accompanying systems for local optimization of production processes, companies have achieved dramatic and very impressive improvements. Inventory turnover rates have been improved and lead times have been cut down significantly resulting in shortened cash-to-cash-cycle-times, improved cash flow, and finally corresponding improvement of the competitive position.

Set up quickly for flexible production

For most concepts and systems, a rich fund of knowledge is available in the form of books, websites, forums, blogs, and by no means least, the services of consultancy companies, all of which companies can use. As mentioned, successes are achieved by implementing the various components of the concepts. Synchronizing processes and controlling inventory lead to a speed-up of production.

Delivery flexibility realized as production flexibility

In order to speak of operational excellence, however, this is only half a step in the right direction. According to the definition outlined above, customer value plays a crucial role, and besides speed, delivered volumes, stability regarding quality, and meeting delivery dates and prices, another key performance parameter, which, inter alia, is expressed in the requirement for JIT and JIS deliveries needs to be considered:

Flexibility

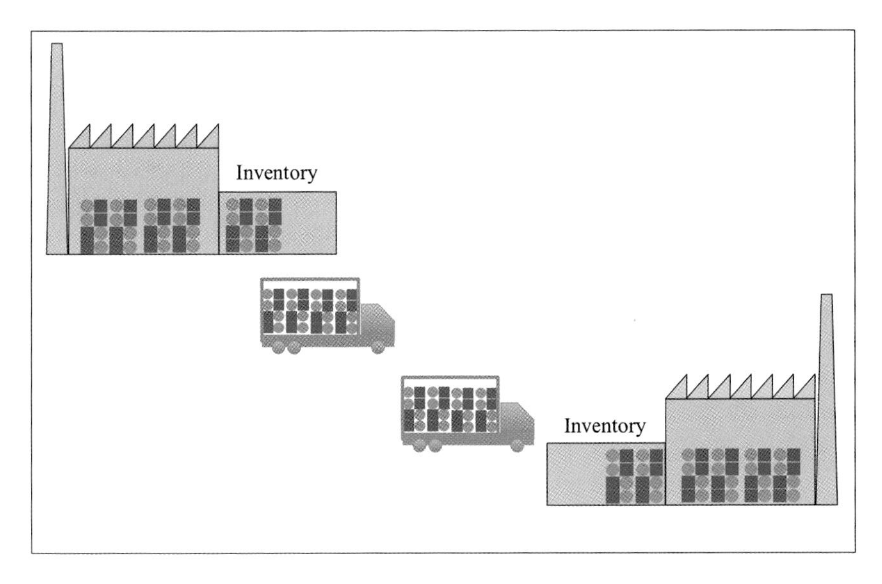

Figure 1-2: Just-in-sequence delivery.

Metaphorically speaking, it could be said that corporate tankers were made ready for speeding up. However, just as before, quick turns are frequently impossible.

In principle, the required JIT/JIS delivery could be achieved by holding the corresponding inventories of finished goods, from which deliveries can be drawn as required. However, this represents a sub-optimal solution at best, because the inventory and production batches associated with it often also lead to a corresponding level of inventory within the production process and thus once again to production being decoupled from the market. In addition, challenging boundary conditions on the capital-demand side cannot be met in this scenario. Thus, for true operational excellence, the required flexibility has to be achieved not as mere delivery flexibility but as genuine production flexibility.

The cornerstones for this are quick changeovers, or

Set-up reduction as the most important cornerstone for operational excellence!

respectively, the reduction of the set-up time. According to the definition above, operational excellence will be achieved by companies that are able to act and react quickly and flexibly and adapt their production system (quickly) to meet client requirements. This is how all the performance parameters defining the customer value proposition are met.

2. SET-UP IN PRODUCTION

"*Any customer can have a car painted any color that he wants so long as it is black.*" Henry Ford

Capacities are occupied, but there is no value creation.

Setting up production resources is first of all an unwanted event. The machine and the people setting it up are two production resources that are occupied, but which cannot be utilized during this time. Costs are incurred but no value is created. Of course, this is nothing new; the godfather of industrial production, Henry Ford, said it a long time ago when he stated that customers could buy the Model T Ford in any color, as long as it was black.

2.1 Changeover in production scheduling

Changeover is often combined with long waiting times and a tiresome production start-up and is counter-productive to efficiency. Bonus systems encourage avoidance of set-up procedures on numerous occasions: Long machine runtimes appear to result in low unit costs and large production batches mean efficient use of the system.
This argumentation is formalized in the calculation logic of economic batch sizes by production planning and control systems.

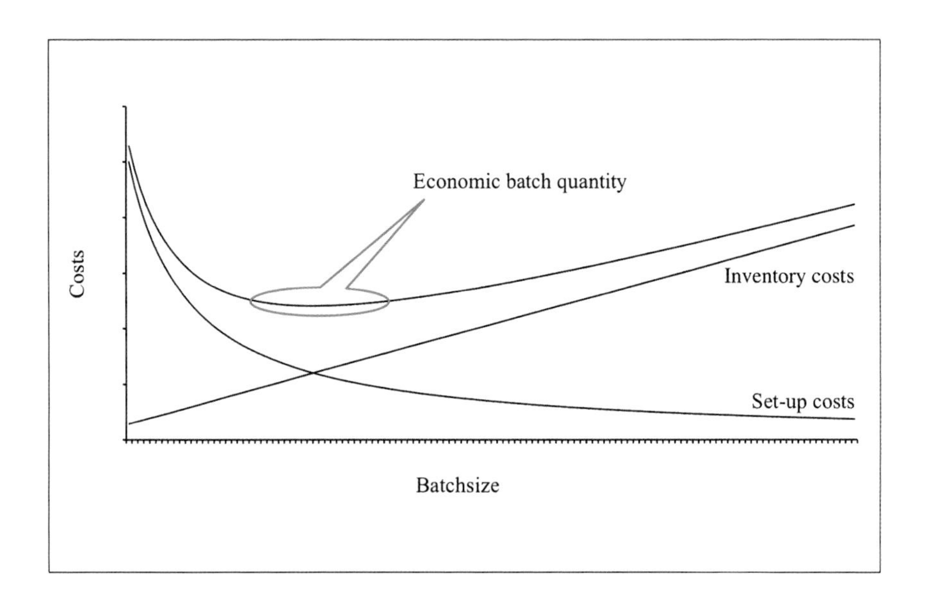

Figure 2-1: Economic batch quantity.

However, this argument fails to shed light on a number of aspects. These large production batches mean increased costs for transport, storage, administration (IT), and the procurement and maintenance of systems for each of these activities. Production faults are often recognized only at a late stage and interrupt the overall scheduling, inventories become obsolete due to product changes and have to be written off, etc. Also, the argument is misleading since large production batches are always produced on the basis of an approximate prediction of the future requirement. The problem here lies in the forecast. The larger the batch, the greater is the uncertainty of any prediction, because it covers an extended period of time. This increases the probability that either far too much has been produced, because the corresponding reserve inventories have been factored in, or far too little. This then applies not to a single product, but to all products, which can cause a certain self-perpetuating effect. The product can no longer be stocked for a variety of

Inventory issues can lead to a downward spiral.

unforeseen reasons, which results in an unscheduled set-up to compensate for the inventory shortage. This again results in all subsequent scheduled production processes being postponed. The sale of products is not interrupted in the meantime, of course. This leads to a range of new inventory issues, which again causes an unscheduled changeover. This again means that all subsequent scheduled production processes will be postponed. Sales of products is not interrupted in the meantime. And so on, and so on.

There is the risk of entering a downward spiral, at the end of which the company finds itself in a situation where delivery dates are not met and customers are dissatisfied. In house, complete chaos reigns and plans are changed on an hourly basis. Finally, costs run out of control, since productivity falls to an all-time low because of constant changes. In addition, the capital circulation time is very high, which does not paint a pretty picture of the company. The operating climate is poor and overtime is the norm.

Chaos, frequent planning changes, dropping productivity figures.

Capital circulation time increases.

Actually, there is a somewhat philosophical problem inherent in the logic that is formalized in the EOQ. These models turn the basic business model upside down to structure the resources of the organization: it uses parts (and therefore finally customers) to keep the machines busy. But that is not what the organization is in the market for.

Production planning system needs to be checked!

2.2 Faulty assumptions

Are long set-up times impossible to change?

The above-mentioned argumentation chain is founded on an incorrect basic assumption – that a system's set-up time is an unchangeable condition.
However, this is not the case, as was demonstrated by *Shingo*. With the analysis techniques described later in this book, which take up Shingo´s SMED-Method and synthesizes it with ideas from modern engineering sciences and the systems engineering approach, it is possible to compress set-ups into very small timeslots.
The introduction of quick changeover concepts allows operations to work with a more real-time production plan which, if used correctly, leads to smaller production batches, lower inventory levels, shorter cycle times, shorter capital circulation times and, finally, more satisfied customers and an achieved competitive edge.

Real-time production.

One of the reasons cited for avoiding set-ups is the tying up of production resources for non-value-added activities. Principally this is a correct statement. However, the context must be assessed accurately: the time that was used hitherto for set-up is usually unnecessary to serve the market. The production time already available is adequate. Therefore, there is not only no need to add the time gained to the available production time — this should be avoided at all costs. To create additional production time and also use it would lead to an increase in existing inventory levels and hence make the situation worse!
However, this may be exactly what happens if a production planning system, which has the efficient use of individual processes as its mere and primary planning objective, is informed that the set-up time has been reduced.

2.3 Objective

A question which still needs answering is what "correct use" looks like. In principle, the motto in production optimization with the goal of operational excellence is:

Don't set up less, but more often!

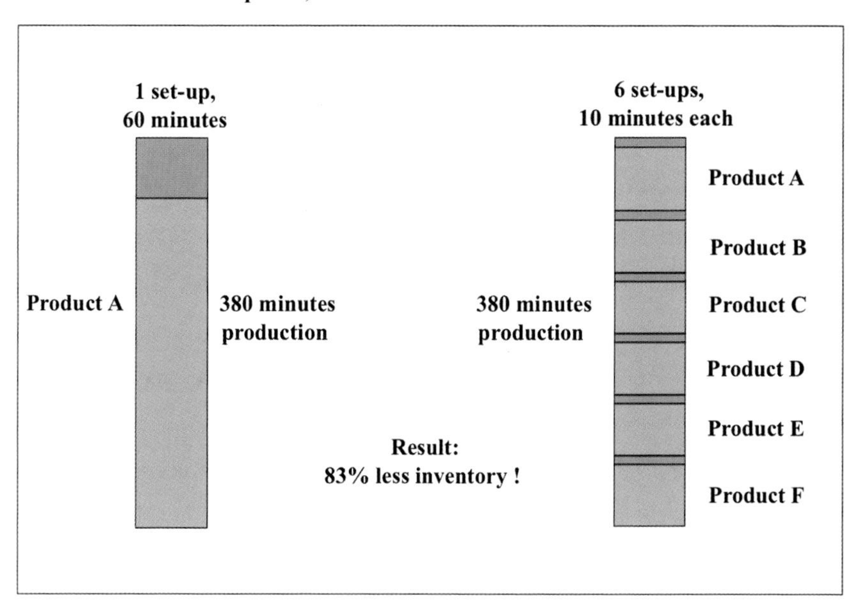

Figure 2-2: Don't set up less, but more often!

This means that time gained through the introduction of quick changeover concepts is used for performing set-ups again. This procedure is shown schematically, although grossly simplified, in the figure. Used properly, this reduces inventory levels, because the production plan is matched constantly to sales reality.

If implemented this way, a company aligns its inner structures with the actual core of its business model: Using resources in order to satisfy market demands.

Setting-up more frequently for flexible production!

This procedure is also used to smooth the production planning. Ideally, as much of each product would always be produced per time unit as is needed to satisfy the market in the given time slot. In Lean literature, the term for this is "Heijunka."

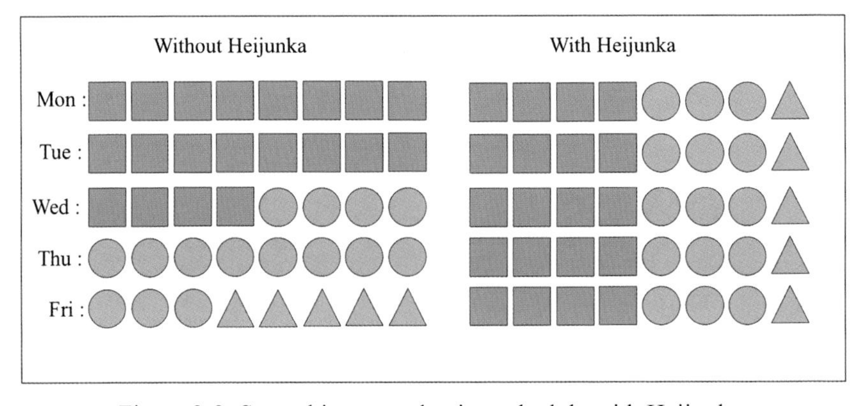

Figure 2-3: Smoothing a production schedule with Heijunka.

Heijunka

Basically the concept is striving for a production mode identical to that of a fast food operation. The production sequence in this case is broadly similar for incoming orders: fries, fries, hamburger, fries, bratwurst, fries, etc. For a more precise description of the Heijunka principle, you can refer to the various books on the subject. Inter alia, recommended are *Learning to See* by Rother and *The Toyota Way* by Liker.

Matching production plan and sales reality.

Hence quick changeovers are a core component in the efforts devoted to placing the Order Penetration Point, OPP, as far upstream as possible and therefore reducing the amount of forecasting in production planning. The diagram depicts the different production strategies schematically, based on the order penetration point. The solid line represents the order-related production component and the dotted line the forecast related. The OPP represents the interface between both value stream components.

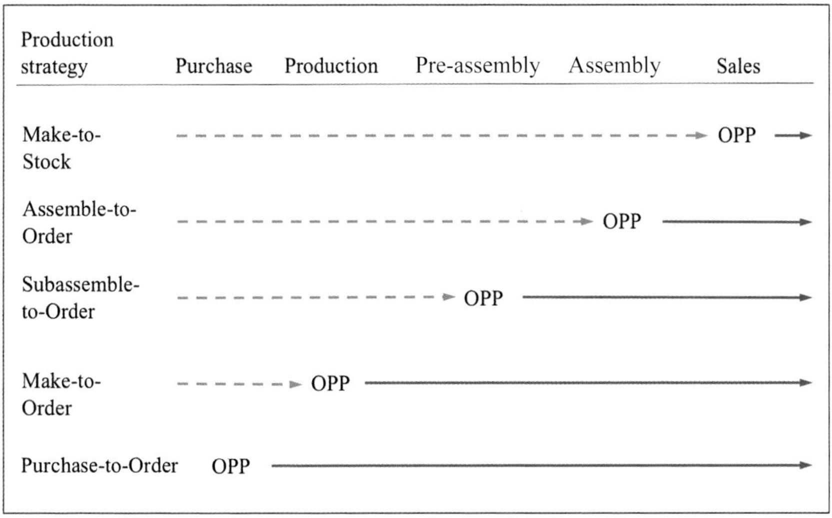

Production strategy	Purchase	Production	Pre-assembly	Assembly	Sales
Make-to-Stock					OPP
Assemble-to-Order				OPP	
Subassemble-to-Order			OPP		
Make-to-Order		OPP			
Purchase-to-Order	OPP				

Figure 2-4: Order penetration point.

2.4 Exception

The exception confirms the rule. What has been written thus far applies to production systems in which products are produced by process chains, with one small but distinct exception.
For bottleneck processes, the five focusing steps of the bottleneck theory apply:

Exception: Bottleneck machines

1. Identify the bottleneck.
2. Exploit the bottleneck to a maximum.
3. Subordinate everything else to the above decision.
4. Elevate the system's constraint.
5. Restart at 1.

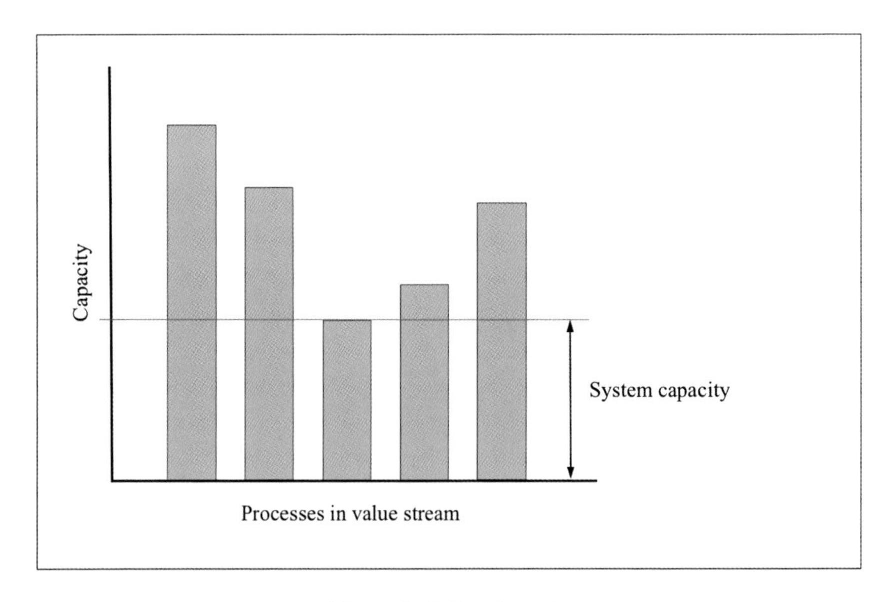

Figure 2-5: Bottleneck.

Steps 2 and 3 describe the need to increase the production capacity and the use of improvement results for this purpose. This can be achieved by moving products away from the bottleneck to other processes — and by assigning the gained set-up time as extra production time.

Before doing so, however, the notion of having a bottleneck needs to be examined very carefully! Most companies these days don`t have real bottlenecks in production, their bottleneck is the market. The feeling of having not enough production capacity in most cases comes from a production schedule that is decoupled from market demand. The effects of what has been described in Section 2.1 may lead to situations where the organization constantly has to keep running behind schedule, leaving the impression that there is a bottleneck issue where there isn´t one. It´s "only" the effect of not being just in time. If that is the case and the decision is made to elevate the systems "constraint" by assigning the gained time as production time, the system will end worse off than ever before. As Russell Ackoff said, "the righter you do the wrong thing, the worse it gets".

Examine situation carefully for fake bottlenecks first!

The objective discussed here in Sections 2.2 – 2.4 will be return in discussions related to changeover costs. There are several widespread ideas concerning this point, one of them being the idea of "lost" production hours. It´s very much the question whether this really applies. In any case, there needs to be carried out a careful analysis of the actual situation to find out the real nature of the cost drivers. In Chapter 12 there will follow a more detailed discussion of the topic.

2.5 Definitions

Having arrived now in the midst of the discussion, there is a need for defining the two core terms *set-up* and *set-up time*. This appears trivial at first glance, but a closer look at the generally disseminated definition reveals that in this case there are ambiguities from time to time. These ambiguities need to be sorted out to have a safe ground for further decisions.

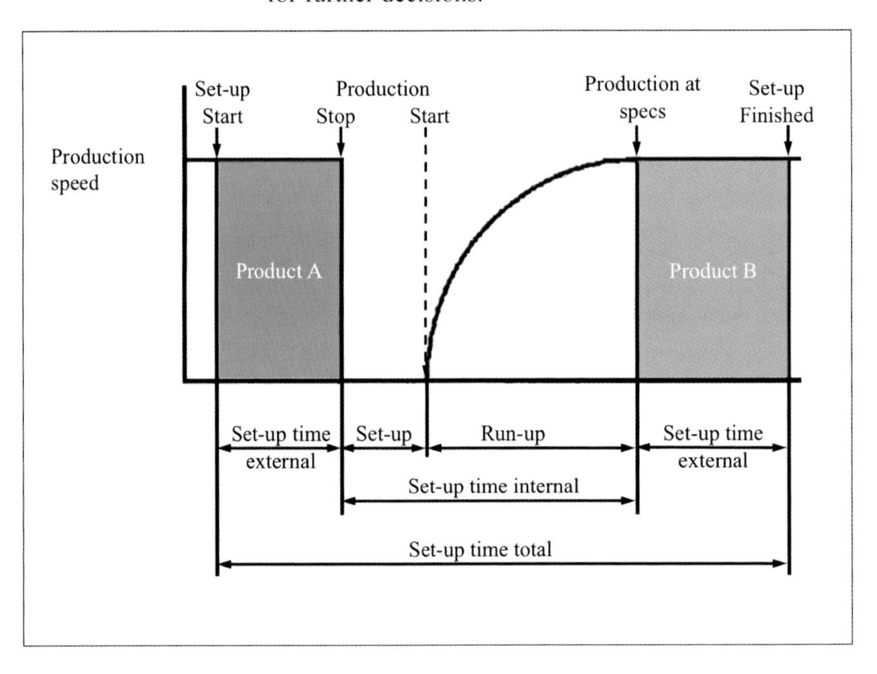

Figure 2-6: Set-up — schematic representation.

2.5.1 Set-up

Set-up = all work steps

Set-up embraces all necessary steps to change a process from one production configuration to another.

This usually includes installing, removing, and conversions of and at tools, units, guides, and fastenings, as well as the creation, installation, and change of programs and running up of the process with the new product, including all setting and adjustment tasks.

2.5.2 Set-up time

Set-up time is the total sum of all individual times taken for these steps.

Set-up time = Time needed for all work steps.

<u>Caution:</u>
Set-up time is *not* the same as the widely accepted and used definition, according to which this is the time between the last product A and the first standard product B.
This is the sum of machine downtime and run-up time, shown in Figure 2-6 as internal set-up time.
The set-up time can be, but does not have to be, the same as this definition. Poorly organized set-ups are often consistent with this definition. The process then corresponds roughly to the schema: machine off — changeover — start machine — production. For a clear definition and for later calculations, the aggregation of man and machine must however be avoided. There needs to be clarity on which resource is occupied for how long and with what. The occupied resources are usually the employee(s) and the machine.

As will become clear in the remainder of the book, when setting up, there are distinctions between work steps being performed externally and internally. In this scenario, there is the question of whether the relevant step shortens the machine runtime or not. Whenever this is not the case, say as in preparatory actions during the production of A, the necessary time according to the broadly accepted definition would not be part of the set-up time and therefore set to "0." However, this is obviously inaccurate, as the employee cannot be involved in other, value-adding, activities elsewhere during the execution of this action; therefore, this time must be included in the set-up time. A distinction is then made between external set-up time and internal set-up time.

3. SYSTEMS THINKING

The methods set out in the remainder of this book are focused strongly on the systems engineering approach, a short explanation of which follows.

3.1 The system

The central object in the system engineering concept is the system. A system is a whole that is defined through its function within a larger entity. Its simplest definition reads like this:

"Every system is always part of a larger system, its surrounding system, and comprises its parts, the sub-systems. It is defined not by the sum of its parts, but by the product of the interactions of the parts."
— R. Ackoff

System: Result as result of component interaction.

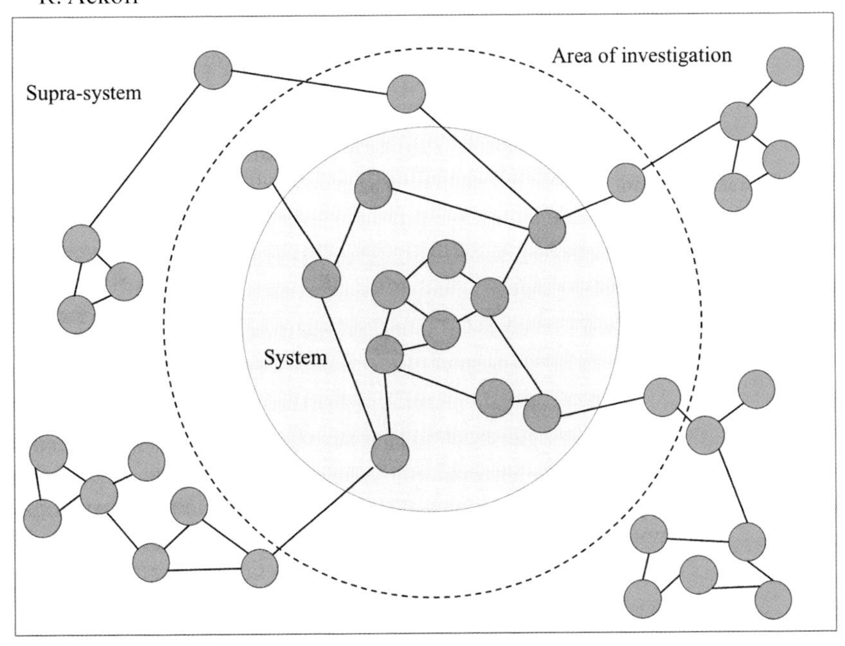

Figure 3-1: System, supra-system, area of research.

3.2 Analysis vs. synthesis

Conventionally, problems are solved analytically.
The analysis process culminates in knowledge about the essence of a whole and its internal dependencies.

The analysis is a process in three steps:

1. The entity is broken down into its parts.

2. Efforts are made to understand how the parts work.

3. Aggregation of the understanding of the parts for a understanding of the whole (which is a synthesis step).

Outside-in, from outline to detail

The synthesis as procedure reverses this process and tries to compile a new whole from the elements which were found with the analysis. Dialectically, the synthesis raises the individual to the general level, the definite to the abstract, and it compiles the manyfold into a single unit.

Synthesis is also a process in three steps:

1. Identify the surrounding system of which the system is a part and it's functionality.

2. Understand how the surrounding system works.

3. This understanding of the surrounding system is disaggregated by identifying the role or function of the system to be explained.

Synthesis and analysis in iteration

For both methods, analysis and synthesis, the steps can or must be performed repeatedly.
The difference is in the method of working. For those

who want to gain more knowledge on the methodology of systems thinking a reference in-depth survey is the book *An Introduction to Systems Engineering* by Rainer Züst, published by Verlag Industrielle Organisation.

3.3 Top-down–outside-in

In systems engineering, a top-down, or outside-in working and thinking methodology is applied as an operationalization of the synthesis process. Ideally, in this scenario, the analysis priorities will be integrated in order to work in an iterative process.
The processing of a question invariably starts with a consideration of the surrounding system and is aligned with the functions or objectives to be achieved.
From this follows a description of the target requirements and the associated boundary conditions for the system and its components. Within this concept then, the work is based on the analytical approaches.

4. PRODUCT PORTFOLIO AND PROCESS

As described in the chapter on systems engineering, the technique for the improvement of changeovers presented in this book, is based on a top-down/outside-in logic. The basic assumption is that each system represents a sub-system of a larger system, with which it is connected accordingly. This surrounding suprasystem represents the starting field in systems engineering.

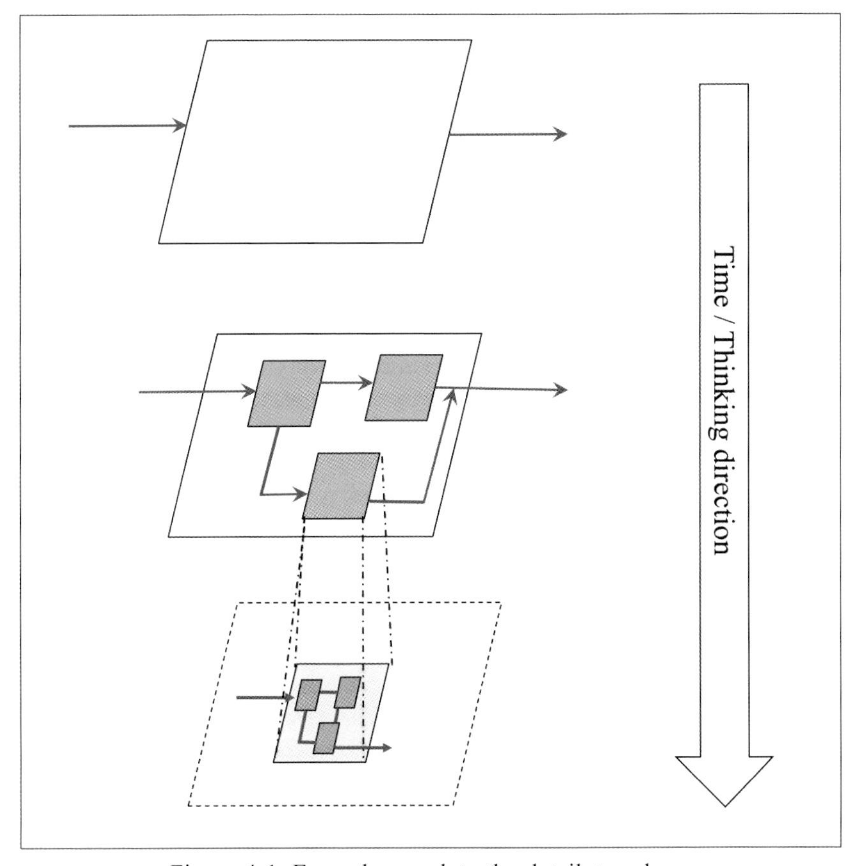

Figure 4-1: From the rough to the detail, top-down.

4.1 Strategy

Initially, when introducing set-up reduction, there are some strategic questions to be asked and answered:

Determining strategy, objectives, and boundary conditions.

- What objectives are to be achieved with the introduction of quick changeovers and under which boundary conditions?
- In what sequence is this to happen?
- And where precisely and within which value streams is the implementation to take place?

A range of further questions arise from these issues, and there are a series of side issues to be addressed. For example, it is essential to first establish which set-ups should be analyzed first:

Which set-up first?

- The short ones to create examples showing the opportunities to introduce one-piece flow production as soon as possible ?
- The long and complicated ones, because the effect on the inventory level is the greatest?
- The simple ones to create employee confidence in the methodology?
- Or maybe set-ups which have the greatest effect on the total inventory storage volume, because space is currently the most critical resource?

Continuous change or continuous improvement?

At the outset, therefore, clarity needs to be achieved concerning the goals to be reached and questions need to be answered concerning what contributions the method must and can deliver at which point.
In general, this requirement applies to all optimization efforts since the target for all optimization programs should be continuous improvement, not merely continuous change.
The necessary clarity can be provided by a decision matrix, by means of which the relevant objectives and boundary figures can be given weighting factors to put them in an overall perspective.

Especially if the company and its employees are at the beginning of the optimization efforts, it may be advisable for tactical reasons to place particular weight on factors related to confidence of the workforce in the methodology. After initial successes are achieved on a permanent basis and the confidence of the workforce is won, the weighting factors can then be moved in the direction of production strategy decision criteria.

4.2 ABC analysis of the product portfolio

If the production system is not subject to absolute restrictions, such as acute shortage of space, which results in an almost natural prioritization of which the workforce no longer has to be convinced, the question is: Which value streams must be optimized, how, and in what sequence?

Customer value as an external and internal parameter

Fundamentally, customer value, both as an external and as an internal parameter, should be defined at the outset. From an internal perspective, the customer is represented in the form of products, which can again be depicted as value streams. This implies, expressed in simple terms, that there are more important and less important customers, and therefore products and thus value streams. Accordingly, there are also more important and less important processes that must be optimized.

How important are the individual value streams?

An ABC analysis of the product portfolio can give a clue to this. There has to be a discussion about which product the company earns most from and which product is most important for the survival of the company (these are not necessarily the same products!). The processes which generate these A-products should receive priority in optimization efforts.

Priority for A-products!

Since most companies these days are operating in high-mix, low-volume environments, the process chain which generates these A-products will normally produce a full range of different products in different variations. This means that a very wide variety of set-ups take place in individual processes (A > B, B > C, C > a, a > A, etc.). This combination of possible set-ups must again be analyzed and structured to find the important product combinations and therefore set-up combinations. Products sold frequently are also set up frequently and efforts should be allocated accordingly.

Frequent products are frequently set up.

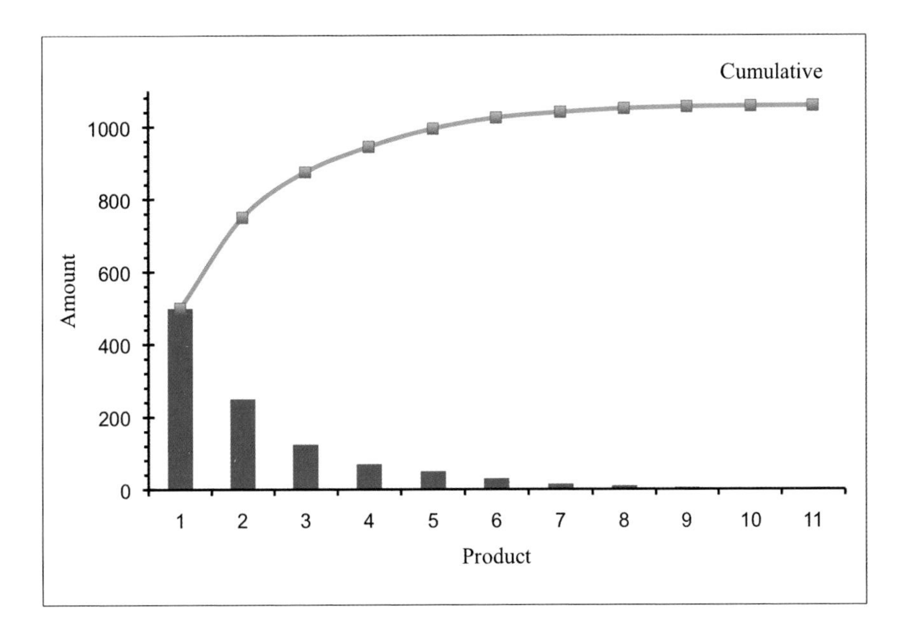

Figure 4-2: ABC analysis.

After the individual set-ups have been prioritized within the relevant value stream, the question of tactics must be answered. Principally, there are possibilities to either optimize a machine and its set-ups completely and then move on to the next, or to optimize step by step at all processes at once.

Optimize in a structured manner!

Depending on this choice, there will be differing requirements for communication with employees.

4.3 Process objective

Functional product analysis

After the most important processes and set-ups in the production system have been determined, the next logical step to be performed is also outside the actual "sub-system set-up."
The question has to be asked: What functionality is to be achieved by the next product component to be set up and could this functionality be achieved in another, simpler way? If there is a similar component or product at another site, by means of which the relevant functionality could be achieved, the evaluation has to be made as to whether this would be a sensible alternative. This applies also to processing: If a corresponding process is already set up elsewhere, it may be more sensible to transport the component or product rather than changeover.

Departmental borders as barriers

The division of companies into functional departments can be problematic in this regard. Potential departmental egoisms and conflicts between departments can lead to the familiar throw-over-the-wall syndrome.

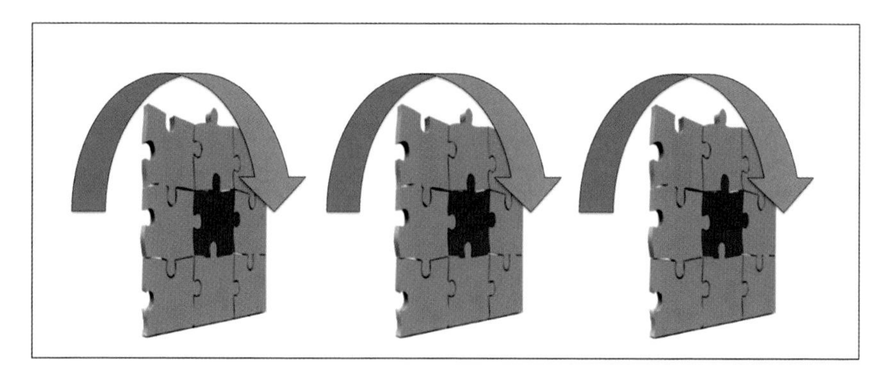

Figure 4-3: Throw-over-the-wall syndrome.

Subsections of problems are then processed to sub-optimal solutions and forwarded to the next department without considering any further boundary conditions, needs, dependencies, and opportunities.
These occurrences should be countered, because they can lead to negative effects, both in everyday operations and related to set-ups. In the area of set-ups, the throw-over-the-wall syndrome takes effect in a completely unique way when set-up time is "reduced" by appointing the performance of certain steps to other shifts. In particular, preparatory steps are often treated this way. It may then be that the individual sectional budget is relieved, however, clearly no contribution is made to total optimization by doing so.

Shifting set-up times is not shortening set-up times!

Engineering, purchasing and production must work hand in hand, in a synchronized manner, to produce the most favorable total solution in products in each case. During engineering, purchasing and sales the possibilities of the production processes must be known and taken into consideration when taking decisions.
If introduction of another version of a component, such as a screw, relieves burden on the purchasing budget, the additionally created production complexity must be evaluated with regard to the question whether the gains really hold in the end when total costs are summed up.
In relation to set-ups, the question that must be asked is, whether the introduction of the further variation leads to additional set-up steps. Also the opposite question needs to be asked: may it be possible to eliminate set-up steps entirely by reducing variations with regards to product engineering? Besides the effects on set-ups, increases in complexity through variation expansions increase the probability of errors and defects. These effects should not be underestimated. To be able to lower the disadvantages described above (or raise the advantages) there is the clear need for interdepartmental communication. Cause–effect relationships do not care about departmental borders.
A special need for additional communication other

Reduce total complexity, avoid sub-optimizing!

Avoiding variations

than the normal may arise in contract manufacturing surroundings. Where contact with the client was formerly exclusively through sales, early involvement of production in discussions and planning can unlock significant potential.

4.4 Standardization to shorten set-up times

Standardizing product engineering, material, processes, geometries, etc., can make a crucial contribution to reducing the overall complexity of the production system and to simplifying set-ups, if not making them completely unnecessary.

Reduce varieties.

A critical concept in this regard is differentiation power.

As long as a component or feature does not have the power to directly differentiate the product from others in the eyes of the customer, the question must be asked whether it is possible to work with standardizations.

Determine differentiation power of features.

This means that, e.g., engineering should on purpose work with dimensional errors and over engineering if this makes a contribution to simplifying the realities of production. An example is the deliberate oversizing of connecting elements such as screws, nuts, and bolts if this eliminates the need for tool changes (besides other effects such as simplifying drawings and machine programs, reducing the number of alternative items to be held in stock with all the accompanying administrative by-products, reducing opportunities for error, etc.).

Standardization by targeted oversizing

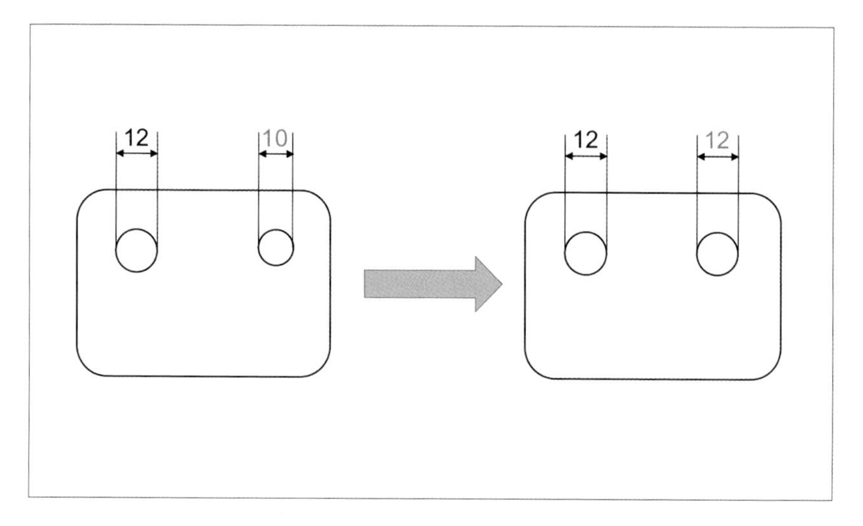

Figure 4-4: Oversizing for reduction of production complexity.

Adapt product design to production possibilities.

Product engineering should ideally be aligned as far as possible with the possibilities of the production system and production resources.

In existing set-ups, therefore, all processes to be performed on the product, and accordingly to be set up, should be questioned at the outset in order not to simplify the unnecessary: Does a process really have to be so precise? Why are so many different drilling diameters needed?

The necessary needs to have priority over the possible.

The possibilities of the production system need to be the central guidelines for the decision processes in surrounding functions.

4.5 Appropriate process

As a last preparatory step, the question to be asked is whether the process, by means of which the desired processing will take place and which is now set up, is actually the right process for this purpose.

The seven wastes described in lean management include the category "Wrong Process/Overprocessing." This is a category that is often difficult to grasp. In the relations described here, it is clarified by the requirement of the "least technical commitment" in production.

Least technical commitment for flexibility reserves

Calculation methods for machine rates based on cost allocation can reveal themselves to be a problem if, for example, they make the boring of a single hole on a simple upright drilling machine more "expensive" than executing the process on a high-tech CNC processing centre. The necessary discussions concerning this must be conducted with the controlling department.
Should it be possible to perform the machining with another, simpler process, the production route should be changed accordingly. Obviously, this must be clarified before starting an optimization of the set-ups.

After this step is performed and the sub-system set-up has been investigated from the outside for opportunities and dependencies, the analyses of the actual work steps of the set-up itself can begin.

5. ELEMENTS OF THE SET-UP

The set-up embraces all necessary steps to change a process from one production configuration to another.

Usually, this covers installing, removing and conversions of tools, units, guides, and fastenings, as well as the creation, call-up, and change of programs for system control and the run-up of the process with the new product, including all setting and adjustment tasks.

5.1 General structure of the changeover

If this general description is structured and ordered, setting up a process can be essentially illustrated as follows:

Tear down product configuration A — Set-up product configuration B — Run-up B

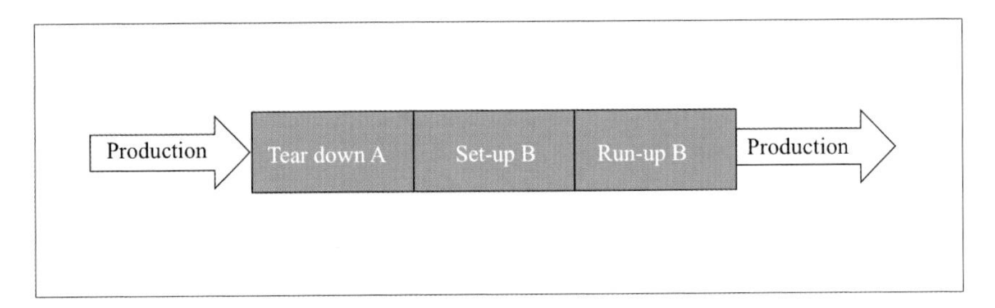

Figure 5-1: Changeover — general structure.

The expanded illustration includes preparatory and subsequent tasks.
Setting up a process then looks like this:

Preparation for the set-up — Tear down product configuration A — Set up product configuration B — Run up B — Postprocessing of the set-up

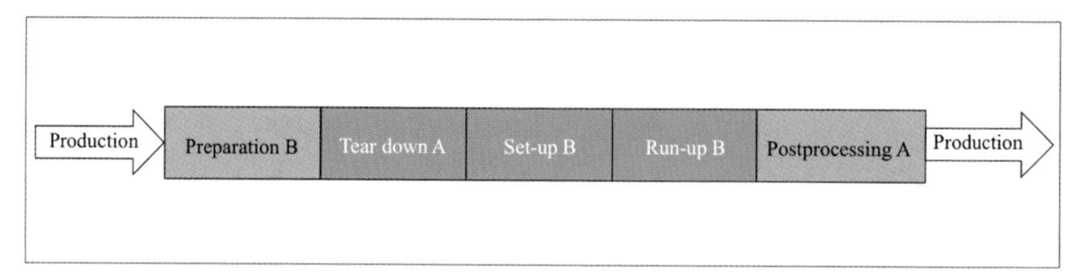

Figure 5-2: Changeover — expanded illustration.

There are various production resources involved in the changeover: the process to be set up itself, one or more employees, and various auxiliary and transport resources.

In the following chapters, this aggregated process will be disaggregated and analyzed step by step to gain understanding about the whole, develop improvement solutions for the parts, and concurrently find solutions to improve the complete process of changeover.

5.2 Areas of the changeover

Setting up and the activities performed to set up a process can be divided in two fundamental areas.
The first area relates to the process hardware, i.e., tools and units, guides and fastenings. The corresponding area in a value stream diagram would be the material flow.
The software represents the second area, i.e., all information in the form of programs, settings, etc. The corresponding area in a value stream diagram would be the information flow.

Isolate areas of the set-up ...

When performing any business process analysis, it is of vital concern to clearly sort out the several areas.
Systems engineering is speaking about dividing the system in "aspect systems." The focal direction is different here, horizontal, then when dividing the "system in subsystems," which is clearly having a vertical focus.

The big pitfall that needs to be avoided comes later, when confusion of areas and categories leads to confusion about the total. (partial-) aggregation earlier then leads to getting stuck in the analysis and losing the oversight. Especially in very complex set-ups, this can mean an end to the whole improvement project.

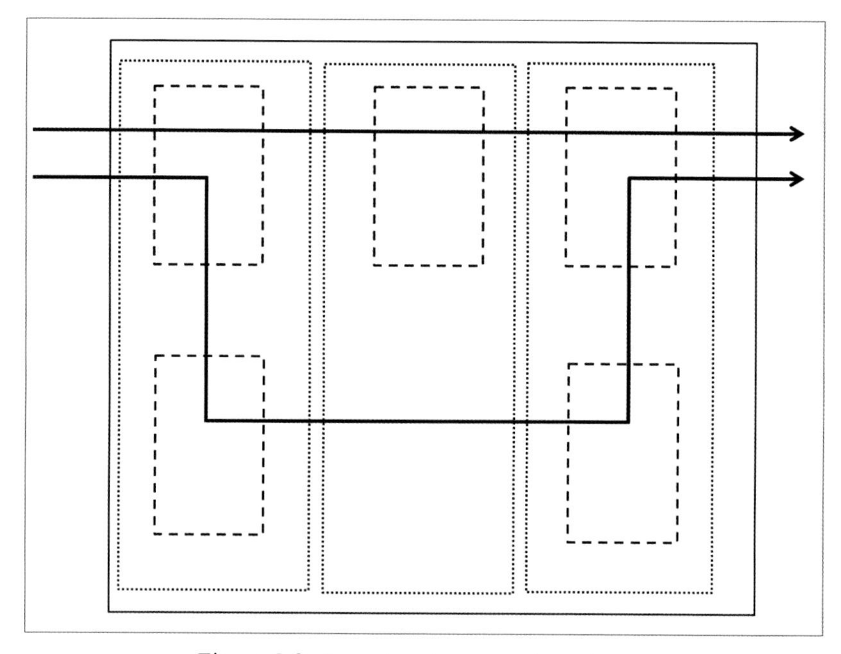

Figure 5-3: Aspect system vs. sub-system.

5.3 Categories

The activities performed in both areas can be grouped and sub-grouped into different sub-areas or categories according to their type. This can therefore mean that an activity can be assigned to a number of different categories.

...and categorize.

This summary of similar activities is useful and relevant, because it gives information about the fundamental character of the set-up. Later on in this book, it will become clear what conclusions can be drawn from the differences between and assignment to sub-areas.

Analysis for structured processing

Frequently occurring sub-areas to be differentiated are:

- Searching
- Movement for transporting items
- Movement without transport
- Installing units
- Removing units
- Assembling units and fastenings
- Dismantling units and fastenings
- Using tools
- Using means of connection
- Settings and adjustments
- Test run/Run-up
- Postprocessing
- Administration

Re-create the category list for every set-up.

By no means, this list does claim to be complete. It is derived from the author's experience and must be generated and adapted for each set-up to be analyzed. This may mean that certain categories do not occur or that other specific categories must be added when a new set-up is analyzed.

5.4 Internal and external execution of tasks

Internal = In-line

External = Off-line

A further, fundamental difference relates to the execution as an internal or external task step and, the need for execution as such.

Activities are described as executed internally whenever the process is stopped. For every task, it needs to be determined whether the activity must be executed internally or if it is an activity that could be executed externally, without having to stop the production process. This is the case for actually all activities that can be performed without the "use" of the process.

Carry out preparations and postprocessing before and after, i.e., externally.

At the beginning of a set-up analysis, the situation is often depicted such that the necessary tasks are executed substantially, if not completely, internally. During a changeover reduction, the steps are structured according to their character. During a changeover reduction the general focus is always on simplifying and shortening all the tasks and steps as far as possible, and in any case organizing them in a way that they can be executed as external steps.

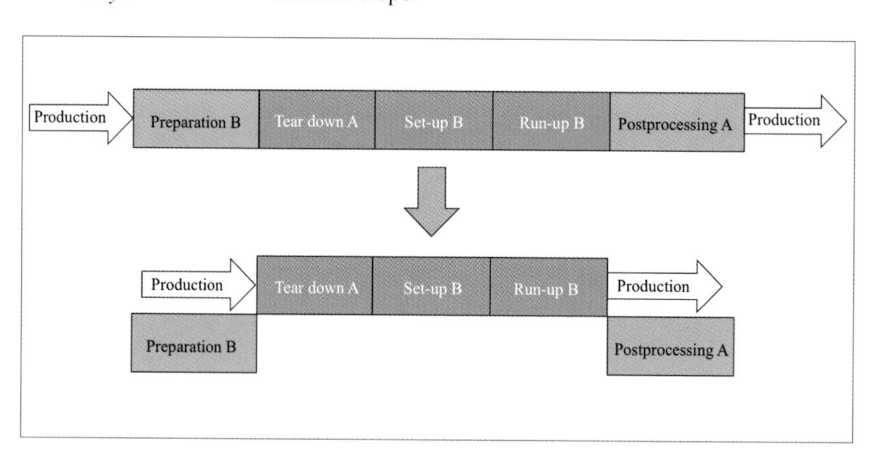

Figure 5-4: Executing external tasks externally.

5.5 Avoiding waste

Waste must be eliminated in all areas and all production areas and processes must be organized to be waste-free as far as possible. With regard to identifiable wasteful activities during a set-up, walking distances play a major role.
They need to be determined both for the existing and later for the newly installed set-ups. Also there should be a check once in a while, how the set-up performs in these regards. One option to analyze the process and visualize the results is to make use of spaghetti diagrams.
The benefit of the spaghetti diagram is mainly in the clarity with which it shows waste by movement and superfluous paths. The impression of the image is substantially stronger than that of the bare numbers.

Visualizing the waste of walking, transport, and moving.

During the process of analyzing the changeover, all valuations and discussions need to be made on the basis of quantifying indicators. Anything else leads to a bare exchange of opinions, in which the majority view or that most forcefully presented is implemented.

Develop indicators.

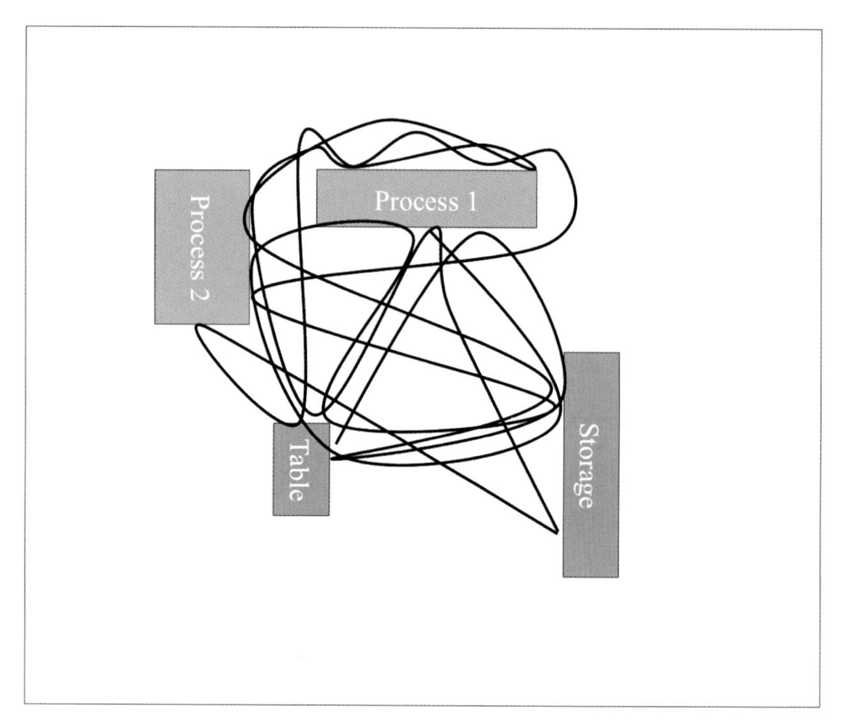

Figure 5-5: Spaghetti diagram visualizing paths.

At this point, however, we need to point out some of the pitfalls along the way of waste identification. Simple identification and elimination of tasks and work steps as wasteful activities, in the author's view, often represents a solution that is too simple.
Steps identified as wasteful must be evaluated, both to establish the extent of the waste and in regard to their value relative to adding or subtracting from the burden on the workers.

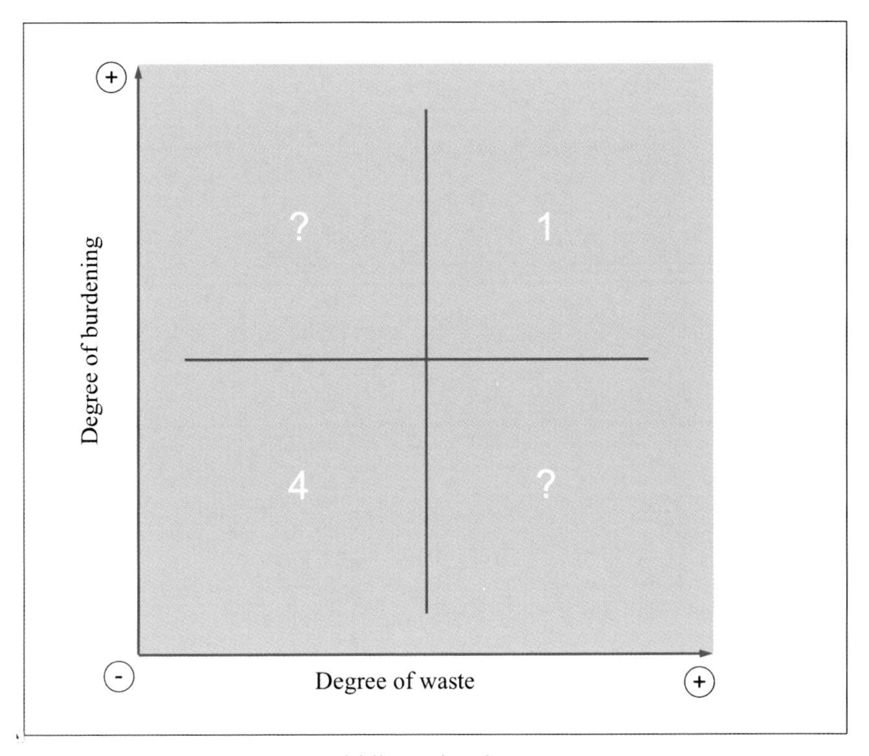

Figure 5-6: Multidimensional waste assessment.

There will always be steps which contribute substantially to relieving the burden on employees, but actually represent waste. These should obviously be kept off-stage in the first instance. Successful improvement plans usually target the most onerous steps, regardless of how trivial the actual waste figure is, and thus create the necessary buy-in. In any case, reducing waste should not lead to an increase on the workers' load. Within the classical lean management literature the concept of Muda is accompanied by the concepts of Muri and Mura as well. It should be treated in an integral manner.

6. ANALYSIS OF A SET-UP

To develop quick changeover concepts, a fundamental analysis of the existing set-up is required after the synthesis step, in which the environmental conditions are clarified as being performed.
The easiest way to perform this analysis is to take a video of the set-up and analyze it. The benefits are obvious: There is no need to interrupt the production planning, the process can be repeated as often as necessary, times are easy to determine using the film time drive, and the film can be stopped at any random point. In particular, with ergonomically demanding positions and postures, the film is clearly superior to the true situation.
Also, the perception of a process through the separation depicted by the monitor is different, which can generate new insights from time to time.

Video analysis as a beneficial tool.

Besides the obvious positive factors of being able to perform a process study from the video, there are however, several aspects that need consideration in order to avoid negative effects.

Depending on the work culture of both the company and the country, some points may be stronger than others.
In any case, you are cautioned to pay close attention to the legal and cultural requirements, challenges, and demands. Especially implicit, cultural factors often need to be taken into account since it is very easy to overlook them.

6.1 Taking the video

Points to pay attention to when producing a film for this sort of analysis:

Respect employee rights!

Include the employees!

Film what actually happened.

- Inform your employees concerning your intention to film a work process and explain your purpose. If there is formal employee representation or an worker council in your company, they must be informed first. In any case, be sure to obey the relevant statutory rules!
- Always respect the employee's image rights! Interpret doubt of any kind as refusal and accept this without further discussion!
- There may be implicit cultural reasons why some person may not want to be filmed. Make sure that you are aware of the culture in general and in particular, the culture of the employee in question.
- The film will be used to analyze the set-up, not the employee. Ensure that this is public knowledge and proceed accordingly. The focus will be on what steps are to be performed, not on how they are performed.
- Avoid the impression of a secret work study. If you want to carry out a work study, then you should make this clear.
- Allow the film to be produced by the relevant employee who is entrusted with and participates in the set-up and, if possible, include him in the process of analysis. This reduces the feeling of being supervised and involves the employees in the improvement.
- Choose a representative set-up and a representative employee.
- Film the set-up. Usually the employee's hands will be located at the center of the action to be filmed.
- Zoom in as far as possible. In so doing, make sure that it is still recognizable what is being

done to which component.

- If the employee leaves the process, it should be obvious where he is going and why. Walk behind him.
- Film the process more than once to avoid actor effects. In particular, if employees are unaccustomed to being filmed at work, the first recordings will only be partially useful.

Be wary of actor effects.

6.2 Video analysis

6.2.1 Analysis sheet

To analyze the video, an Excel spreadsheet similar to that in the following figure can be used.

Analyze the film in steps.

The sheet offers the basic functionalities that are needed to separate the work steps, assign them to their corresponding categories, and analyze these individually.

I will be happy to make the worksheet available on request. If you want to receive a sheet, please don't hesitate to contact me via email at karstenherr@inmatech.de.

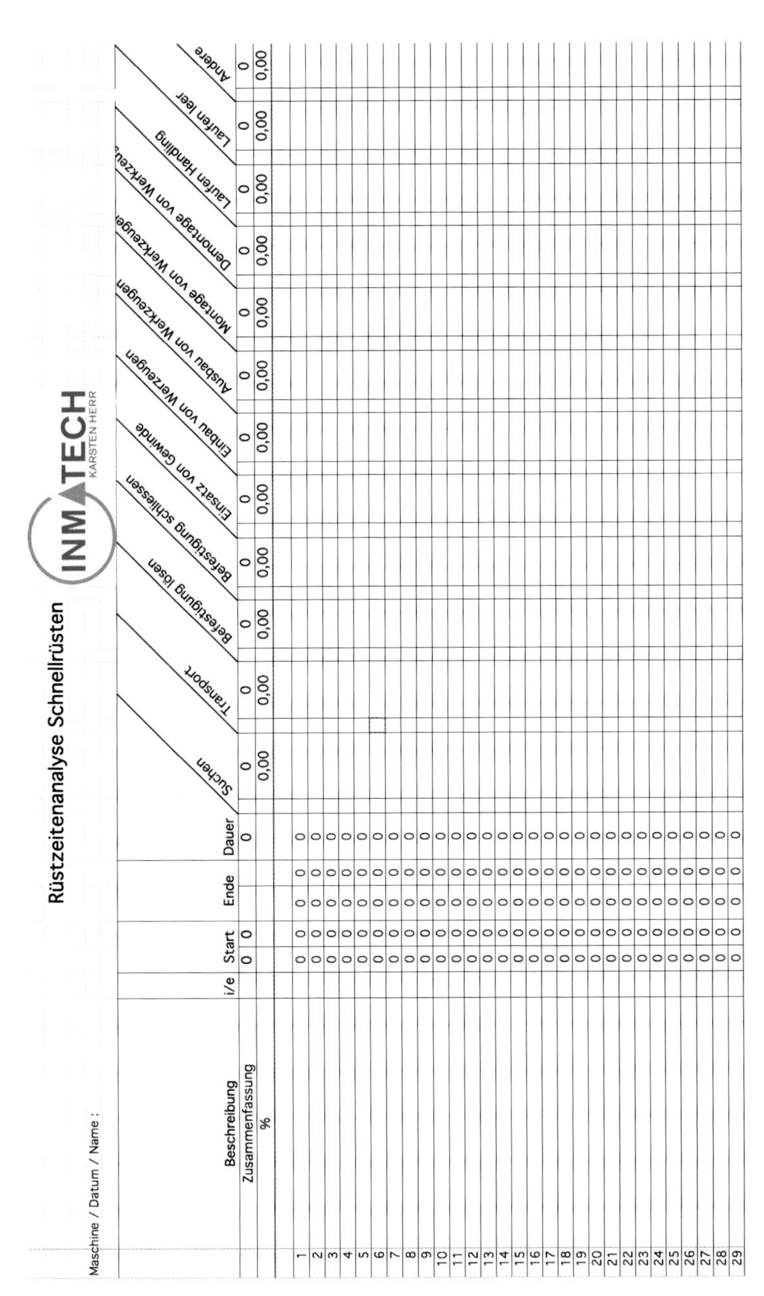

Rüstzeitenanalyse Schnellrüsten

INMATECH KARSTEN HERR

Maschine / Datum / Name :

	Beschreibung	i/e	Start		Ende		Dauer	Suchen	Transport	Befestigung lösen	Befestigung schliessen	Einsatz von Gewinde	Einbau von Werzeugen	Ausbau von Werkzeugen	Montage von Werkzeuge	Demontage von Werkzeu	Laufen Handling	Laufen leer	Andere
	Zusammenfassung		0	0			0	0	0	0	0	0	0	0	0	0	0	0	0
	%							0,00	0,00	0,00	0,00	0,00	0,00	0,00	0,00	0,00	0,00	0,00	0,00
1			0	0	0	0	0												
2			0	0	0	0	0												
3			0	0	0	0	0												
4			0	0	0	0	0												
5			0	0	0	0	0												
6			0	0	0	0	0												
7			0	0	0	0	0												
8			0	0	0	0	0												
9			0	0	0	0	0												
10			0	0	0	0	0												
11			0	0	0	0	0												
12			0	0	0	0	0												
13			0	0	0	0	0												
14			0	0	0	0	0												
15			0	0	0	0	0												
16			0	0	0	0	0												
17			0	0	0	0	0												
18			0	0	0	0	0												
19			0	0	0	0	0												
20			0	0	0	0	0												
21			0	0	0	0	0												
22			0	0	0	0	0												
23			0	0	0	0	0												
24			0	0	0	0	0												
25			0	0	0	0	0												
26			0	0	0	0	0												
27			0	0	0	0	0												
28			0	0	0	0	0												
29			0	0	0	0	0												

Figure 6-1: Analysis sheet.

6.2.2 Procedure

If you are new to this sort of process analysis, it is recommended that you analyze the film in a number of steps or cycles, starting with a rather general focus in the beginning and later detailing the subsequent steps. Otherwise the task can become too complex due to information overload.
Once the necessary routine has been developed with these analysis methods, shortcut working methods will be generated automatically.

In the first step, the film will be shown and it will be determined which general dependencies and categories exist. If nothing is transported, it is not very useful to handle a transport category.

Work in detail.

In the following cycles, the set-up will be analyzed and described step by step. The analysis of the work steps becomes increasingly detailed.

The steps and tasks will be described, and it will be determined whether it is performed internally or externally (in the beginning this will very often be internal) and it will be decided whether the step or task *could* be performed externally.
After making externally performable steps external, the work steps need to be shortened, beginning with the internal steps. This corresponds to the classic SMED methodology procedure.

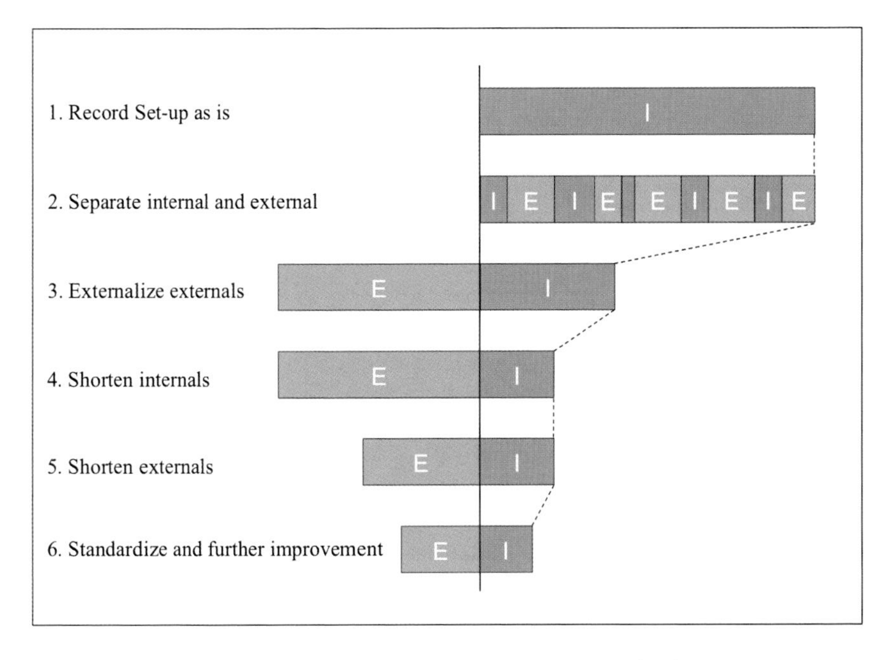

Figure 6-2: Analysis and step-by-step optimization.

For analysis and categorization, the start time and the relevant end time of the work step are listed. This end time becomes the start time of the next step automatically. Identified components are assigned to the relevant categories. It is possible that a component is assigned to several categories, since not all categories that are relevant and interesting are also always independent. It is mainly the case if, for example, a separate record is made as to the role of threaded connections. This then generates an assignment to the categories Assembling/ Disassembling and Threads.

Describe every single action.

The times of the individual categories are then summed and expressed in relation to the total time.

Depending on requirements, it may be necessary to describe the work steps up to such a detailed level that for the conclusion only a list of verbs is left describing

every single action of which the process step is made.
At some point, further detailing may not result in additional benefits, but too low a level of detailing is a problem when working out solutions.

Particularly when first performing this sort of analysis, there is from time to time an inclination to first work in very great detail, followed by assuming that it is *too* detailed and re-aggregating individual actions.
The aggregation then no longer records what actually happened in detail and it accordingly becomes impossible to work out a solution.

Avoid aggregations

The highest aggregation level would be achieved if this said: " The process was set up and it lasted x hours." This is correct, but it represents nothing more than the starting point, at which the opportunity was seen to read this book or to concentrate on the topic in detail.

The reason for this effect to appear is the earlier mentioned loss of oversight due to not clearly defining the several areas and categories, describing the sub-aspect systems in which the whole process can be decomposed.

6.3 Data analysis

The data won provides information about the character of the set-up. It may be interesting to search for other set-ups showing the same structural patterns. If, e.g., a repeating pattern is long times for searching, this is directing toward a possible solution: do a 5S campaign, If there is a frequent pattern showing lots of time required for transport, think about relocation. Also, data analysis is a tool that is relatively easy to handle for making it clear to employees where the focus should be when assessing processes and work sequences. The strength of the method lies in the fact that once explicitly identified dependencies are perceived differently. People see what they know.

Recognize structures.

A further point dependent on this and not to be underestimated: Because people see what they know, this is mainly what they see. Without a detailed analysis, one tends to deal with certain issues preferentially, although an analysis would possibly reveal that other points are more important. Conclusions can now be drawn from individual categories:

- Is there a lot of transportation?
- Are there frequent searches?
- What proportion of the respective set-up do settings and trial runs make up?
- How controlled are the set-ups and processes?

Compare patterns.

6.4 Consequences

With the analysis technique described above, individual work content is processed in stages. In addition, every step is treated on its own merits, without considering further dependencies. In order to reach a conclusive solution concept, however, attention must be paid to naturally determined systemic dependencies. The individual work steps do not exist for themselves, but only in combination with all others. This internal organization of work steps must be given consideration, because significant improvement potential can be hidden here.

Do not assess steps independently of each other.

Because the approach followed in this book is not so much about optimizing existing elements, but rather about redesigning of complete set-ups process organization issues will be covered in detail later on in Chapter 10 only. It would be a waste to first put energy into the design of a new process organization and then afterward change the individual steps, which would then make the results of the reorganization efforts completely or partially obsolete again.

Don't optimize first and eliminate later!

The approach taken in the remainder of the book assumes that the following sequence applies when processing the individual steps:

Eliminate, Simplify, Organize!

7. FUNCTION, PROCESS FORCE, AND INTERFACE ANALYSIS

Analysis of the functional properties of machine elements

After the sub-system set-up has been analyzed for its various dependencies and links with its "environment" in order to clarify any fundamental changes in the production process, as well as in the product as an alternative solution, the work steps of the actual set-up can now be looked at. In this chapter, therefore, the individual machine elements will be analyzed step by step and the design of the elements will be looked into for improvements of their functional properties.

The set-up is sub-divided into the (main) activity areas set-up, teardown, and adjustment. During set-up and teardown, various mounting and breakdown tasks, as well as assembling and dismantling of machine units and tools and other machine elements, are performed. The objective of the work done in each case is to change the machine-unit combination to the next product.

Implement DFMA as "Design for Use."

When engineering machines and processes, the machine manufacturer makes a variety of efforts to overcome in-house barriers and raise internal efficiency. In so doing, the throw-over-the-wall syndrome, mentioned previously, is countered by various methods, such as Design for Manufacturing and Assembly (DFMA). The core of the methodology is the engineering of machine elements that are as simple as possible to produce and assemble.

It is frequently observed that this methodology is used with a purely internal focus. In other words, a machine has been designed that is easy to build for the manufacturer but not necessarily easy to handle for the user. DFMA must be combined with an external focus, which describes the objectives and boundary conditions for use ("Design for Use"), to generate genuinely usable designs.

A particular problem is the fact that additional requirements related to product alternatives are often placed on machines and processes while they are running, meaning that new units and tools are designed again and again. This multiplicity of alternatives is difficult to foresee in the original machine design. When designing a machine or a process, both producer and client should therefore take sufficient time for analysis of the requirements.

The individual machine elements must support the quick changeover objective in their design.

7.1 Design of machine elements

From the perspective of setting up the machine quickly, various demands are placed on the design of individual machine elements regarding the required properties.
In order to set up the elements as quickly as possible, these must be designed to facilitate easy handling. *Easy* in this regard means:

Design machine elements to be low-weight and small.

Handling and transport can be performed by one person without the need of conveyor belts or lifting gears. Quick to fix without aids.

This boundary condition sets the requirements of designing small and low-weight elements. Lighter components have the advantage of not only being easier to transport, but also allowing easier positioning.

7.2 Function and process force analysis

In order to take into account weight and geometry needs, function and process force analyses must take place to support decisions regarding material selection and element design. The conventional construction material in mechanical engineering is steel. The properties of this material are well known. It is easy to machine, has excellent mechanical strength properties and is very resistant. However, it has the crucial disadvantage that due to its relatively high specific weight, very it often leads to very heavy components.

Carry out function and process force analysis.

At this point, function and process force analysis support clarification of what functions the component should fulfill and exactly by which part of the component geometry this function is fulfilled.

- ➭ What function must be fulfilled?
- ➭ About how many versions are there?
- ➭ How large and of what type are the forces?
- ➭ Which mechanical stresses arise?
- ➭ Where to and by which part of the component geometry are these forces diverted?

7.3 Material choice

The answers to these questions will produce a differentiated requirement picture, which represents the basis for choosing the material according to the relevant material properties.

Adapt material choice to the strength requirements.

Process force analysis also delivers information on the size and direction of the individual stresses.
Where large mechanical forces must be diverted or absorbed, a design using steel will continue to be the most preferential.
However, where this is not the case, the question is whether materials of lower densities can be used to reduce the weight of the component.
Frequently, component analysis shows that it is possible to work with material combinations. The result may be components of which parts still must be designed in steel, because they are exposed to corresponding stresses, but of which the rest can consist of another material, because it has other functions, such as bridging a space, for which no additional mechanical properties are required. Light alloys and plastics in this case offer versatile options.

Avoid solid material

Sheet metal for light-weight tools

Another approach, which is also based on the function and process force analysis, is to avoid or reduce solid materials when designing parts, or later weight reduction by selective drilling out of components. A variation on this is the choice of alternative design technologies, that produces lighter components, such as a sheet metal design.

Process force analysis also delivers information on which parts of the forces must be diverted in exactly which directions. The term *degree of freedom* has to be introduced in this context. First of all, this term will be defined.

7.4 Degree of freedom

The space in which a part is located is described by three axes X,Y, and Z. The position of each part can be described with the relevant three axis coordinates. Also, a part can change position relative to its axes. These movement options are described as degrees of freedom.

Degrees of freedom as movement opportunities

Each body also has two degrees of freedom per axis:

- rotation degree of freedom
- translation degree of freedom

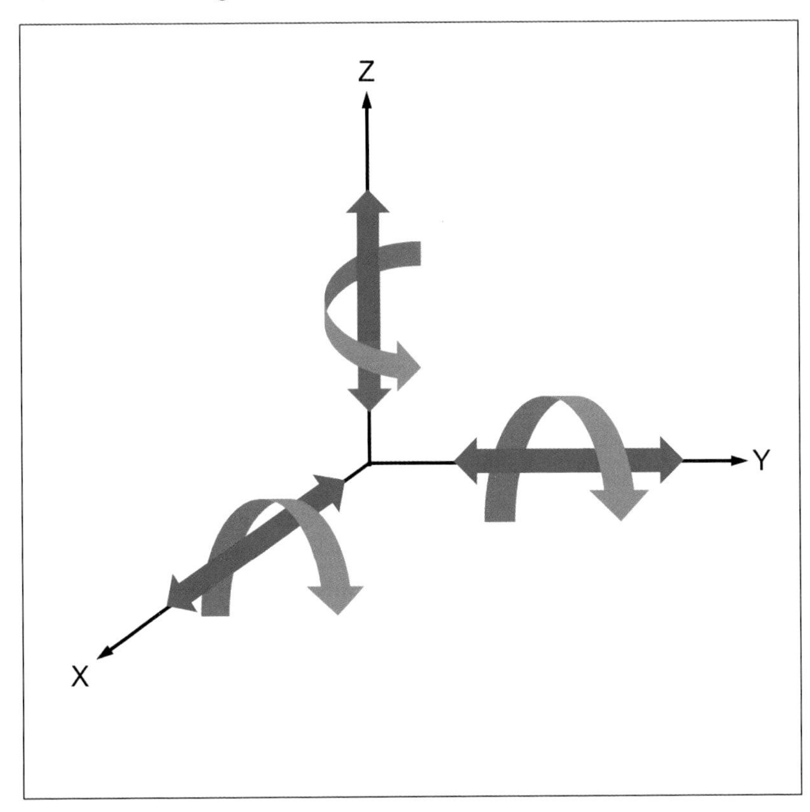

Figure 7-1: Degrees of freedom.

Fastening = Suppressing degrees of freedom

It can therefore rotate around an axis and move along the axis, respectively. Three axes multiplied by two degrees of freedom per axis produces six degrees of freedom per body. If all these degrees of freedom are suppressed, the body is fixed.
The meaning of fastening is preventing movement of the component by suppressing the unwanted degrees of freedom.

Process force analysis for all axes

Process forces are broken down for analysis into each of their three individual axis components. The individual components are then analyzed for their effect on the respective degrees of freedom of the machine element, in order to determine precisely in which directions which parts of the forces are diverted. This information is then the basis for designing the machine element and its fastening.
Process force analysis thus describes the distribution of the force components and their effect on the six degrees of freedom of the machine element.

7.5 Process forces and counterforces

Determine counter-forces.

Process force analysis makes clear which forces and stresses arise and how the corresponding counterforces must be designed. For this, the size, direction, and type of process forces are crucial. The fastenings must be appropriate for delivering the corresponding counterforces. In relation to the type of force, it must be noted at this point that premature decisions on a certain type of fastening must be avoided. In particular, vibrations and designs vulnerable to vibration require precise analysis!

After the theoretical framework is clarified, the various implementation options are discussed.

7.6 Form-fit before force-fit

Principally, form-fit connections take precedence over force-fit connections when designing components and machines. In implementation, this means that stops or rests and slide-in units should be used whenever possible. These form-fit connections are accompanied by a side effect — a positioning element is connected with them concurrently. Detailed explanations of the relevance of positioning follow in Chapter 9.

Stops as form-fit positioning elements

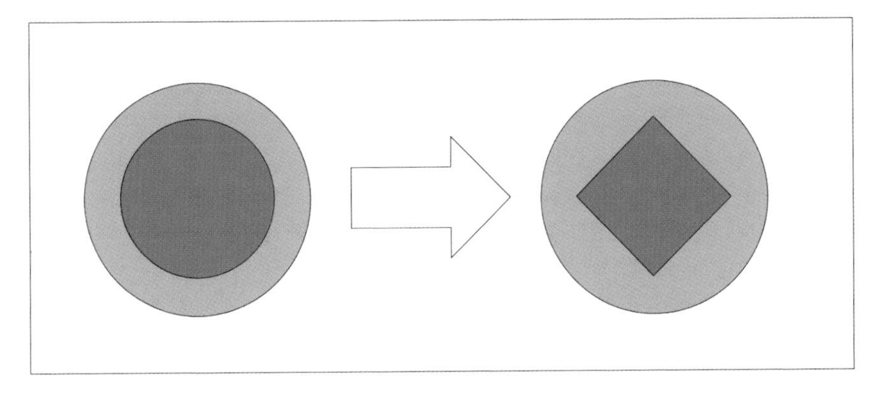

Figure 7-2: Form-fit (right) before force-fit (left).

Form-fit connections are usually produced significantly faster than force-fit, in which the component is brought to the right position in the machine and then fastened.
No mechanical operations are needed and maintenance problems disappear. Also, accidental or independent loosening, and connections that are impossible to loosen, e.g., with screwed connections that are too tight, are nearly non-existent with form-fit connections.

Strive for form-fit.

The main critique with the connection is not the force-fit, but the way the force is produced and the positioning, which is usually not realized directly with a force-fit.
Relatively quick force-fit connections can be produced if these are based on quickly operable mechanisms, such as

toggle clamps in combination with positioning by means of, e.g., a stop.

Realize force-fits with quick connections.

Similarly fast connections based on force-fit can be produced by using magnet or vacuum systems. Magnetic clamping tables are frequently used in the metal industry, vacuum clamping is found frequently in the timber trade (furniture and kitchen construction, etc.) and in both industries, mainly in the area of milling and cutting.
Here, a glance beyond the borders of the own industry is rewarded with new perspectives and insights.

7.7 Diversion of process forces

Frequently, mechanical stresses such as process forces are diverted via the connection elements. This should be avoided since it demands designs of the connectors which are incompatible with fast set-ups. Instead, the main process force is preferably diverted via a form-fit of geometric design, such as a stop. The form-fit topic will be discussed later in this book in more detail in fastening theory.

After the component design or adaptation with reference to the functional analysis has taken place, the next key step can be taken — interface design analysis.

7.8. Interface analysis

In order to reach the core element of the analysis and engineering technique described here, the term "interface" needs to be introduced:

The interface represents the surface or line along which the movable object (unit, tool, guide, etc.) and the machine are interconnected.

At the focal point of interest are questions regarding the possibilities of geometry adjustment and the options that are available when designing the interface. The options are

Scrutinize interfaces

- ↳ Matching the machine to the unit
- ↳ Matching the unit to the machine
- ↳ Using inserts

Cassette principle

In order to achieve really fast changeovers, the design and positioning of the interface must be questioned. Experience shows that interface standardization, next to the question of alternative methods of fastening, offers the greatest time-saving possibilities.
Standardization of the interface follows the objective of working with a "audiocassette principle."

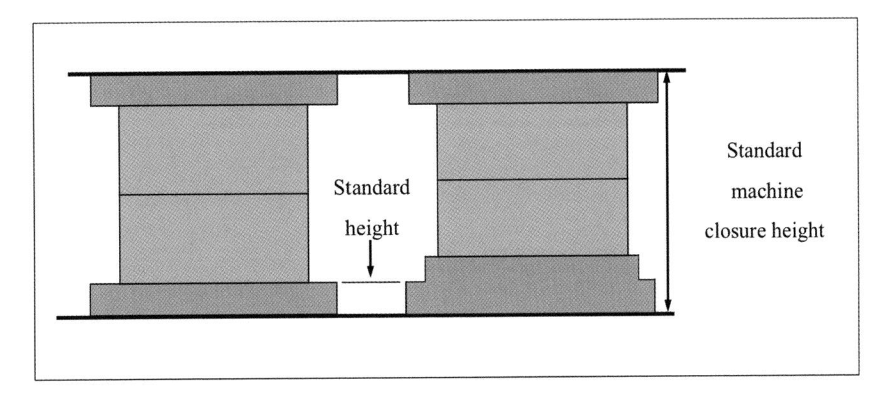

Figure 7-3: Cassette principle.

Standardize external geometry.

The external geometry (cassette housing) is standardized, whereas the functional geometry (cassette tape length) is flexible. Such a modular process and unit construction, with machine elements and interfaces following standardized designs, reduces the overall complexity of the process, reduces the opportunities for error, and simplifies the execution of the process since no machine settings are needed.

Inserts as an alternative

Working with inserts represents an intermediate step to a standardized interface. Bridging the gap resulting from missing geometry avoids having to change the machine settings. However, the introduction of new loose elements should only be chosen when it is impossible to match the exchange element geometry. Introducing individual components with new degrees of freedom first of all increases the complexity of the process and should therefore be avoided.

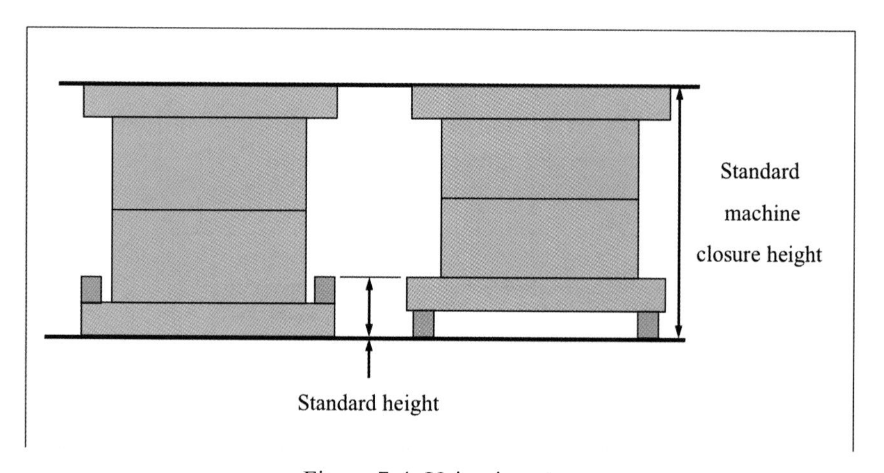

Figure 7-4: Using inserts.

The basic questions for analysis and design of quick changeover systems therefore are:

Interface analysis

Where is the realized interface?
Where is the functional interface?
Where is the ideal interface for the set-up?

7.8.1 Implemented interface

The term "realized interface" is self-explanatory. It is the geometry along which the units and tools are fixed to the machine and on which the corresponding assembly and disassembly work steps are executed. They are the result of the original machine design.

7.8.2 Functional interface

Functional interface: What must actually be changed

The term "functional interface" describes the geometry along which the actual changes would have to take place in order to change the functionality. Often this is not the same as the implemented interface.

The concept can be illustrated with the example of a pit stop from car races, which will be considered later in more detail:
When tires are changed, the interface implemented is between the entire wheel (i.e., the rim including tires) and axle. The wheel is separated from the car at the axle and exchanged as a complete module.
However, the functional interface would be between the tires and the wheel rim, because only the tires really need to be changed. Principally there is no need to exchange the rim.

7.8.3 Ideal interface

The question to be answered is where the interface should ideally be.
In theory, this question is quite easy to answer: Ideally, the interface is set at the highest possible aggregation level. This means that in each case there is a machine set-up for each process. However, this "dedicated equipment" path does not represent a realistic option for most production operations.

There cannot actually be a simple standard answer to the question of the ideal interface. Rather, a range of considerations and tips follow, to answer the question for the process in question.

7.9 Function integration

Basically, all machine elements and units should be designed so that internal assembly and dismantling tasks of the items themselves are avoided as much as possible. Each action to be performed adds cost and slows down the set-up.

Integrate functions, avoid assembly at elements and units

To achieve this goal, the opportunity needs to be clarified as to whether the same function for producing different products can be combined in a single component/ tool. When combining, the focus is then on producing frequently produced alternatives with the tool, because these combinations occur most frequently as set-ups.

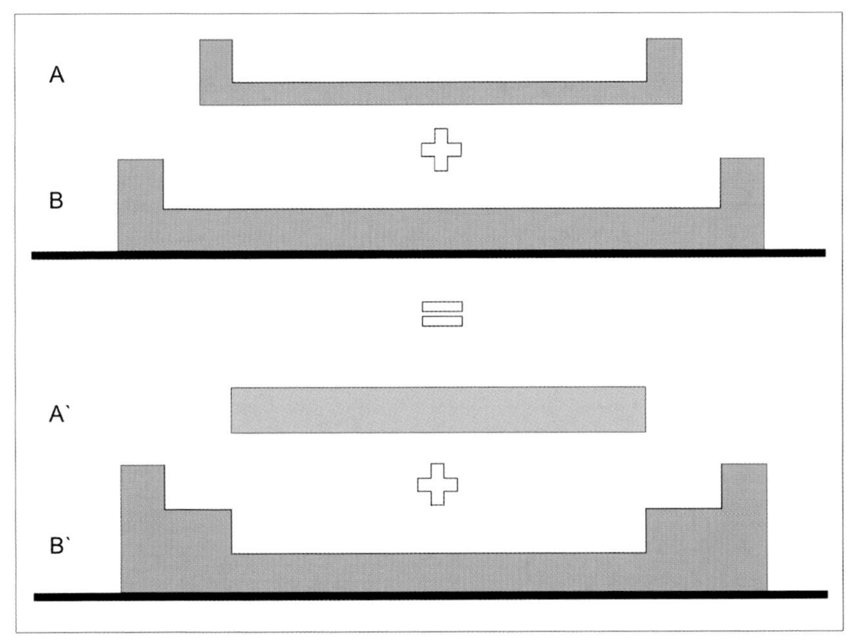

Figure 7-5: Combining tools.

Combine versions of tools

The objective is implemented, for example, when machines are working with standard equipment and tools. Successful companies rely on using only standard tools and designing the relevant product engineering based on the possibilities these tools offer. It can also be wise to use tools that are actually known to be "wrong" (because too large/too small/too high value), if these are already set up in the machine, and thereby a tool change can be avoided. When combining functions in a single machine element, the question must be asked as to whether it is essential that the function must occur as before, what problems this brings with it for production of other versions, and whether another type of production might not be more beneficial.

Standardization by machining with the "wrong" tool

The ideal solution can also require a not insignificant redesign of the component. This point was discussed in Chapter 4.

Do not couple elements to one another.

Generally as part of function integration the consideration needs to be taken whether components can be structured "on one another" in a modular fashion. In combination with a Pareto analysis of individual production volumes, appropriate nesting can lead to the complete avoidance of certain set-ups. The change from A to B and back again to A would no longer comprise the sequence: tear down A, set up B, tear down B, set up A, ... , but rather ...set up B, tear down B, set up B, ... — a reduced sequence. A would then remain completely in the machine

Build components on top of each other

If the different production functions are spread over a number of machine elements and not all elements have to be exchanged with each set-up, you should avoid coupling the elements to each other. Instead, independent fastening of all elements is sensible. Coupling elements in such a case leads to additional assembly and dismantling tasks, which should in any event be avoided. It also needs to be investigated whether elements that are not needed in the next production run have a disruptive influence on production. If this is not the case, dismantling can be avoided.

Leave tools in the process.

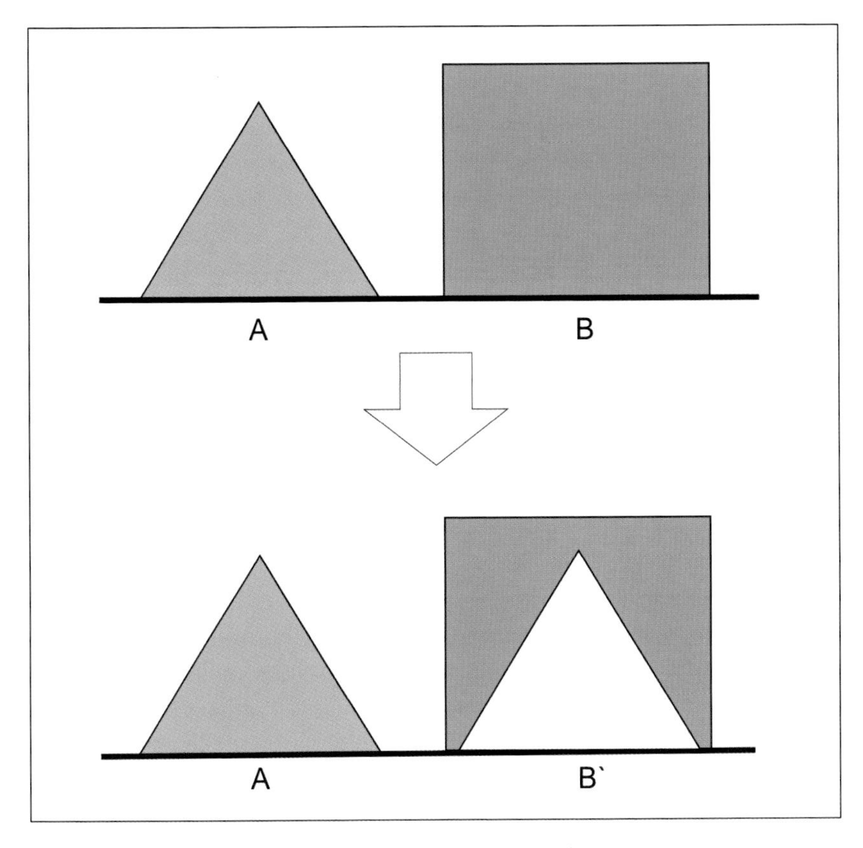

Figure 7-6: Modular design of tools and components.

The main question when designing machine elements should be:

How must the element be designed so that it can remain inside the machine?

Carry out work steps on tools and units as external actions.

Despite every effort to avoid such tasks, it can be necessary to perform actions on machine elements that need to be changed themselves. However, these should be organized such that they can be carried out as external actions. As an example of this, zero point clamping systems for external mechanizing of work pieces, or tool setting devices for preparing tools to be exchanged can be mentioned.

However, these options can be considered only as a second-best possibility. Solutions which result in elimination of actions receive absolute priority.
This is equally applicable to machine parts: Ideally, function integration makes individual machine elements superfluous.

7.10 Designing machine elements and their interfaces

In summary, it can be established that investigating the design of machine elements regarding the materials chosen, the function combinations implemented, and their interface design can release significant potential for simplifying and shortening set-ups.
With the requirement to focus on the facilitation of quick set-ups, machine elements, tools, and units are designed differently than is conventionally done when the focus is first and foremost on simple production.
Working with dedicated equipment is ideal for quick changeovers. Where this is impossible, the design of the machine elements and components to be changed must be rethought. Function and process force analyses give information about needs and options for design. The interface should be selected at the highest possible aggregate level in the hierarchical structure Machine > Module > Assembly > Part. The boundary conditions that have to be taken into consideration are that interfaces must be standardized, in order to minimize the complexity of set-ups and that the components need to be designed ideally regarding their ease of handling (weight, dimensions, etc.).
Functions should be combined in as few components as possible to allow a one-size-fits-(almost)-all system.

THE PROPOSED STEPS IN THIS CHAPTER CAN MEAN A CHANGE OF THE DESIGN OF MACHINE PARTS AND TOOLS. YOU SHOULD NEVER CARRY OUT SUCH CHANGES WITHOUT CONSULTATION OF AN ENGINEER!

IN ANY CASE YOU ARE SOLELY RESPONISBLE FOR YOUR ACTIONS !!

CAUTION !!

8. MECHANICAL FASTENING

After the individual machine elements were analyzed step by step and the design of the elements for improvements regarding their functional properties were investigated in the previous chapter, this chapter approaches the different actions for fastening these elements.
It is clear from analysis of set-ups that a large part of the time needed is used for operating fastening components. Accordingly, there is corresponding savings potential to be realized in the field of fastening elements.

Large part of set-up time for working with fastenings

8.1 Mechanical fastening theory

To be able to explore the various alternative fastening components and methods and assess them according to their potential, it is necessary to consider some general points of attention concerning mechanical fastening theory.

As introduction, some theoretical context is illustrated in addition to the term "degree of freedom" explained previously in Section 7.3.1.

8.1.1 Newton's axioms

In 1687, Isaac Newton formulated his three principles of motion, which are known as Newton's axioms, laws of motion, Newton's principles, or Newton's laws.

Newton's axiom as the basic laws of motion

1. A body remains in a state of rest or uniform translation, unless it is forced to change its state by externally applied forces.

2. A body's change of position is proportional to the effect of the externally applied force and takes place in the same straight line as that in which that

force acts.

3. Forces always act in pairs. If a body A exercises a force on another body B (action), then an equivalent but opposite force is applied by body B to body A (reaction).

These laws form the foundation of classic mechanical science. It is one of the fundamental subjects in mechanical engineering, formulating the equilibrium conditions.

The equilibrium conditions state that a body remains at rest if the sum of all forces and moments acting on it are zero.

8.1.2 Static certainty

Requirement of static certainty

Number of bearing reactions = number of degrees of freedom

The mechanical fastening theory results in the requirement that static certainty must be realized when fastening a component. Designs are then determined to be statically certain if the number of their bearing reactions (support conditions) equals the number of their degrees of freedom and the bearing reactions can be calculated from the external loading with the equilibrium conditions alone.

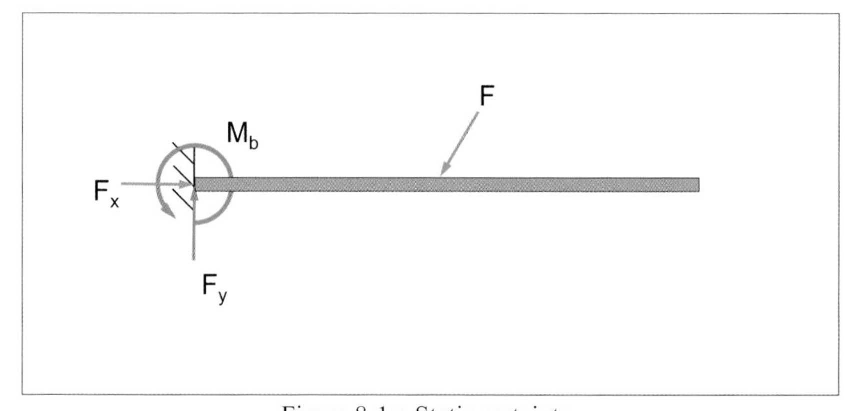

Figure 8-1a: Static certainty.

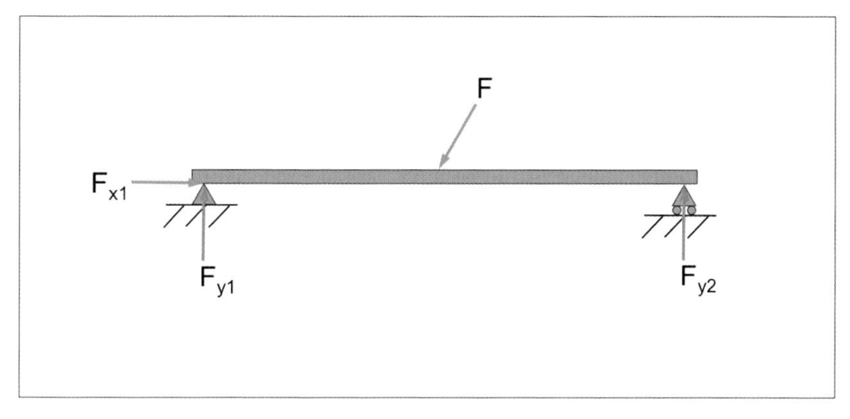

Figure 8-1b: Static certainty.

Scrutinize connection points

In practice, this means that there is a need to work with the minimum number of connection points. Instead, components are often connected with their environment at too many connection points, which not only results in static uncertainty, but also demands too much time operating all these connections.

Before starting to remove screws and nuts and bolts, take into consideration again the warning mentioned at the end of Section 7.10:

DO NOT CHANGE PARTS OF A MACHINE WITHOUT CONSULTING AN ENGINEER!

Some fastening elements may lead to a situation of static uncertainty, but are still necessary, because they may be there to prevent deformation of the machine part that is connected with them. It can be necessary to make the machine part stronger in order to remove fastening elements!

8.1.3 Summary of mechanical fastening theory

As simple as possible, but not simpler

From the summary explained in this chapter, combined with the knowledge from the process forces analysis and Einstein's famous phrase, according to which things should be done as simply as possible but not simpler, three principles for the realization of fastenings can be derived:

Limit degrees of freedom only where necessary

1. The degrees of freedom of a part are to be suppressed by means of a fastening, if a counterforce needs to be realized, directed at a process force or the weight force.

2. The degrees of freedom of a component do not require any suppression if no corresponding process force is present.

3. A component is completely fixed if each of the six degrees of freedom is suppressed precisely once.

8.2 Connections

8.2.1 Threaded connections

Now that the theory of mechanical fastening principles has been presented, the various options for realizing fastenings and connections and their respective implications for quick changeovers will be illustrated.

The method typically chosen for fastening machine components is the screw, or another connection solution based on threads, such as nuts, bolts, threaded rods, etc. These connections belong to the classic repertoire of machine constructors. They are popular, because they are relatively cheap and well known, which makes application simple.

Regarding changeovers, however, threaded connections are conceivably the worst of all possible solutions!

Avoid threaded connections!

Threaded connections have a whole list of crucial disadvantages, the most important of which are listed here:

- It takes a relatively long time to realize the connection. The insertion of the screw requires many turns, which takes unnecessary time, since only the final fraction of rotation provide the actual fixing effect.
- Screws (and nuts) are small, loose elements which can be lost or misplaced during the set-up.
- It is very rare for manufacturers to work with a standard size, which makes searching and sorting necessary.

Threaded connections obstruct quick changeovers.

- They can usually only be loosened and tightened with tools (which again represent loose elements that in turn usually are not standardized; see the above problem).
- Screws can sit firmly and cannot be loosened.

- The closing force is relatively undefined, which leads to tight turning.
- When loosening or tightening, the thread can become damaged or destroyed.
- Wear takes place both on the thread, and at the head of the screw.
- It can take a long time to get the screw inserted and the thread can wear out from the insertion process.

The listing could be continued and clarifies handling problems.
As a consequence of the mentioned effects, often a major part of the time used for changeovers is occupied by the operation of threaded connections!

8.2.2 Friction forces

Friction as basis for a force-fit connection

Fastenings based on friction represent a special type of force-fit connection. The so-called Coulomb friction describes the effect that dry friction causes between two bodies based on the normal force (FN) of the weight of the upper body and a coefficient of friction.

In equation form: $FR \leq f(\mu) * FN$

Depending on the material and surface properties, the coefficient of friction f or μ varies between approximately 0 and approximately infinity.

The values for the most given material combinations can be found in any book of tables.

For the combination of steel surfaces the value typically lies between 0.08 and 0.25.

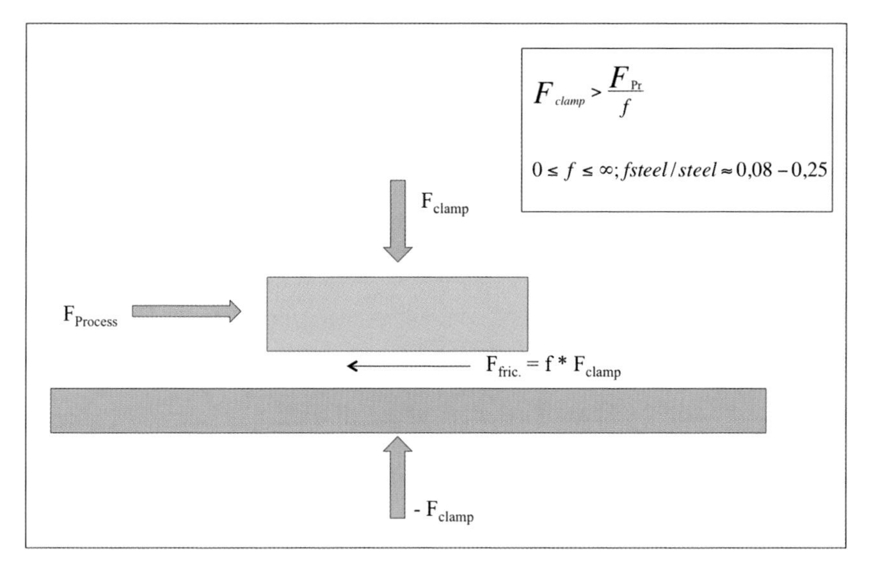

Figure 8-2: Friction.

Realize friction for a force-fit by clamping.

In machines, components are frequently fixed on the basis of this friction force. To achieve functionality, a frictional force must be generated, that in every case is higher than the process force acting against it. To generate this force, it is necessary to apply a commensurately large normal force. Because coefficients of friction of most material combinations used in machines have values very much lower than 1, the result is that very large normal forces must be applied. For an overview of coefficients of friction, refer to the relevant standard works and table books.

Of course the requirement for form-fit before force-fit applies, but in principle a connection based on frictional forces represents an interesting option for connecting components and machine elements, as long as the corresponding force is not based on a threaded connection, but provided by appropriate alternative fastening elements.

8.2.3 Fastening

For quick changeovers of the machine or process elements that need to be exchanged, alternative connection and fastening solutions need to be applied.
These solutions must be consistent with the demands resulting from fastening theory and the illustrated dependencies on process forces and interfaces.
In summary, the following rules can be formulated concerning fastenings and for selection and use of appropriate fasteners:

Fastening globally

Form-fit whenever possible.

- Seek form-fit before force-fit.
- Divert main stress or process force if possible via form-fitting elements.
- If possible, connect parts with hinges on one side only and fold away or push back instead of remove them completely from the process.
- If elements are not interfering with the next production run, consider leaving them inside the process.

Fasteners

Avoid loose elements.

- Not tapped
    - If this is not feasible, threads as short as possible and work with snaplock nuts, etc.
- Use fasteners that can be operated without the use of tools.
    - If this is not feasible , standardize tools and make use of power tools.
- Implement single-handed operation.
- Tailor closing forces to stress profile.
    - With multiple fastening: Standardization, avoidance of alternatives.
- If possible, do not design as loose elements;

produce fixed anchoring in the process or

- ➷ Realize as complete connection element; avoid of assembling and disassembling of fastening mechanisms.
- ➷ Use power-driven connections (hydraulic, pneumatic, magnetic, and vacuum clamping).
- ➷ Leave non-interfering elements in the process.

Sources for alternative fasteners

Clamping equipment manufacturers such as Erwin Halder KG, Kipp KG, Otto Ganter GmbH & Co. KG, Andreas Maier GmbH (AMF), and PM Bearings B.V. produce an extensive range of fasteners which can be used to quickly realize connections.

8.3 Alternative elements for connection and fastening

Fasteners can roughly be sub-divided into two groups, those which suppress the degrees of freedom based on form-fit and those that do so by force.
Locking washers and similar elements do not comply 100% with these categories.

8.3.1 Form-fit

Locking rods and bolts, indexing plungers, quarter-turn closures, bayonet closures, etc., for example, are based on form-fit.

On the following pages, some of the most widely used examples are given.

Most of the elements can be purchased in a wide range of sizes and in different materials.

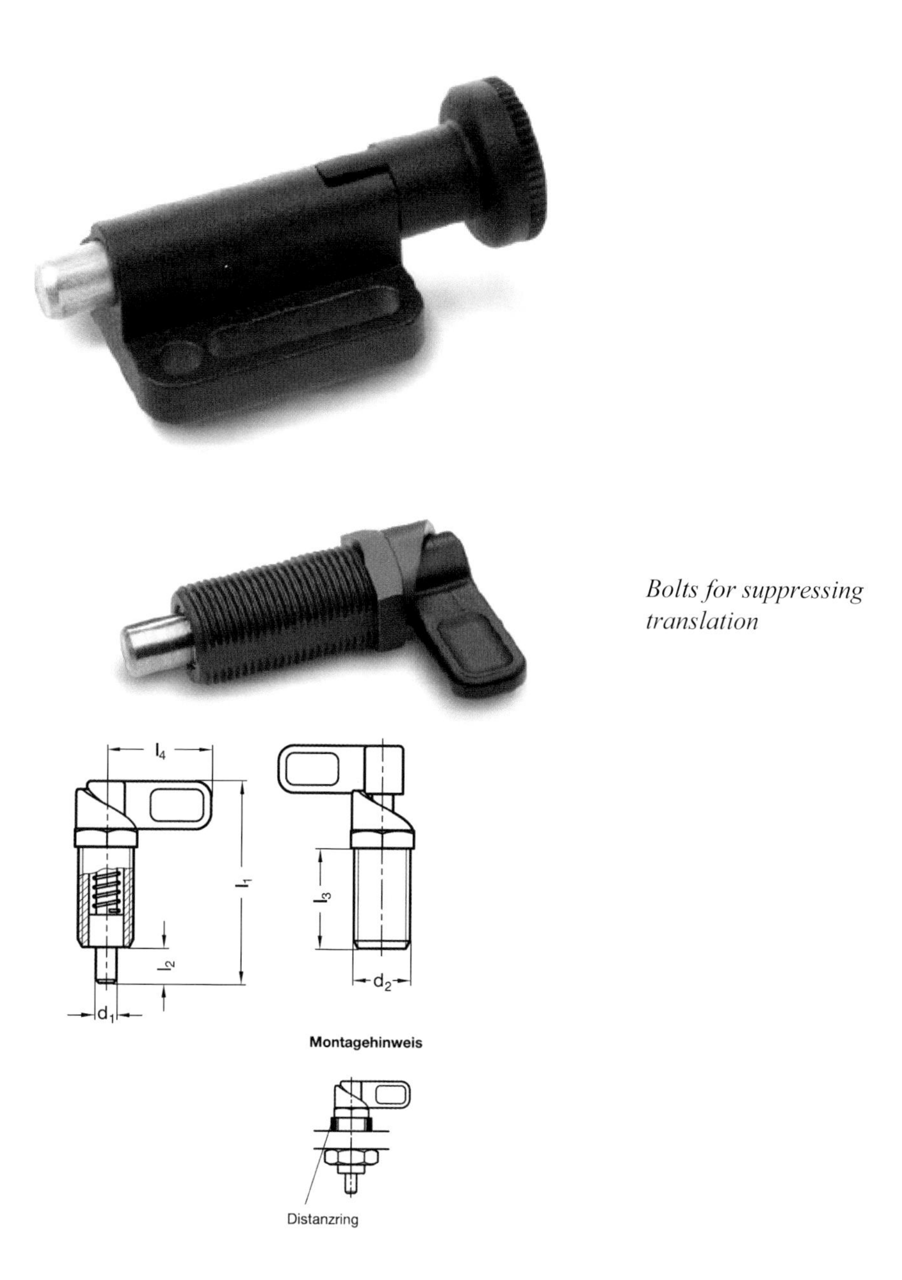

Bolts for suppressing translation

Figure 8-3: Rods and bolts with and without locking.

Locking balls are spring loaded.

Kugeln verriegelt
(90 ° versetzt gezeichnet)

Markierung für W. Nr. 1.4542 (GN 113.8)

Kugeln entriegelt

Druckbolzen

Plug and rest bolts with and without safety mechanisms.

Application Diagram

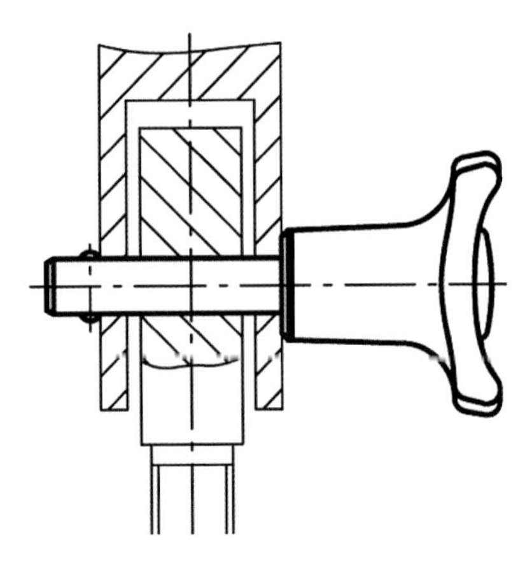

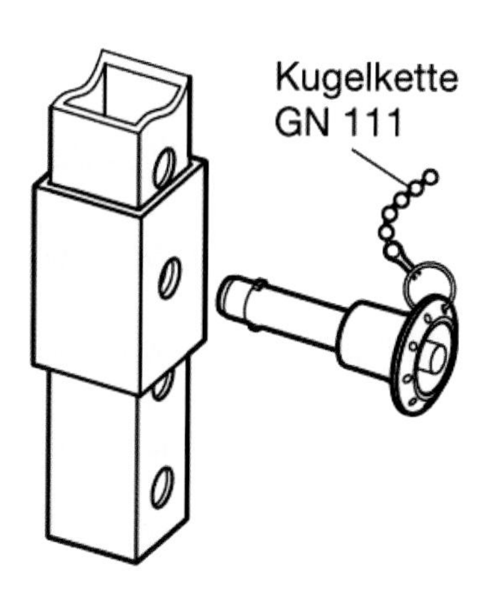

Do not design plug bolts as loose elements.

Figure 8-4: Plug-in bolts with and without axial locking.

Plugs, plunger pins and bolts are offered in various materials and surface finishes and dimensions. Both locked and unlocked spring-tensioned versions are available.

Various forms of bayonet closure represent a further alternative. They are commercially available in the form of quarter-turn screws; in constructions they can be produced both in circular and in flat geometries.

In combination with eccentric cam levers as introduced later, they can be a very effective way of connection.

Use bayonet in combination with clamping closures

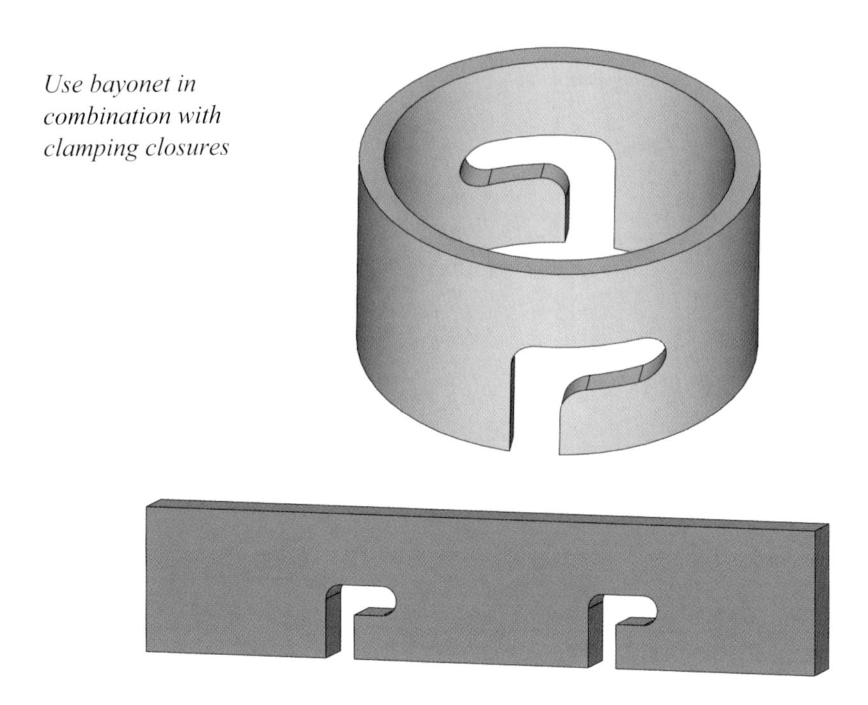

Figure 8-5: Bayonet closure.

8.3.2 Force-fit

Connecting elements which functionally rely on the application forces are, for example, Snaplock nuts, spring plungers, clamping and tensioning devices, magnetic and vacuum systems, wing nuts, cam levers, wedges, buckles and hooks, etc.

The function of tilt and locking washers is based, precisely as with the familiar pear-shaped opening, on an interplay with bolts or screws. The disadvantageous elements of the screw are minimized extensively since it is only necessary to loosen the screw. There is no need for turning it out all the way, also it will not get lost or confused with others, since it remains in place.

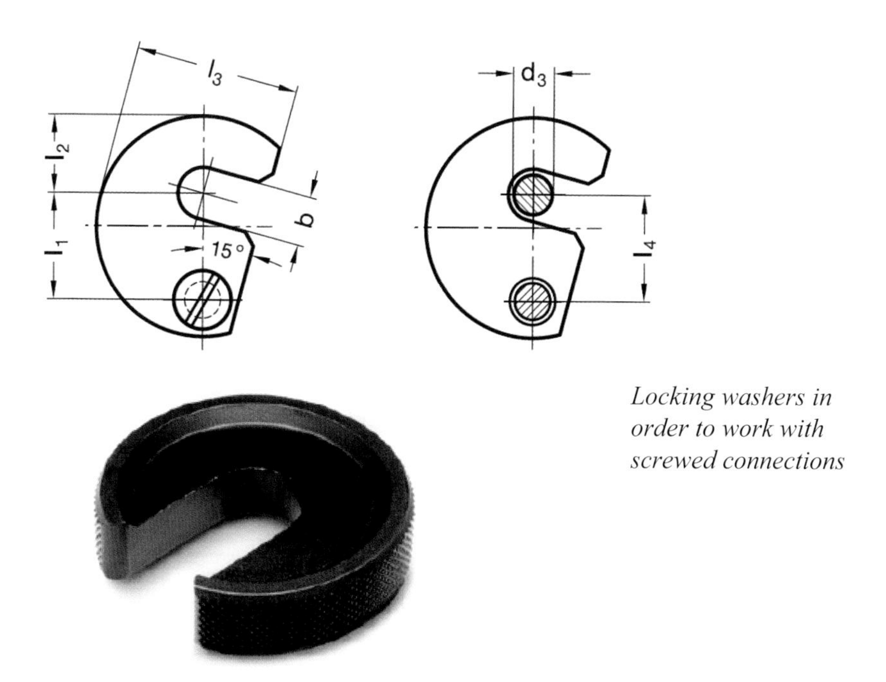

Locking washers in order to work with screwed connections

Figures 8-6: Tilt and locking washers, pear-shaped hole

To meet quick changeover requirements properly, these elements must be combined with screws that do not require tools. Should this be impossible, the tool needed should be anchored on site to prevent searching, etc. Wing nuts/screws represent the most familiar example of zero-tool usable, thread-based connections. Clamping levers, star-grip screws, and similar elements represent appropriate alternatives.

Avoid tools!

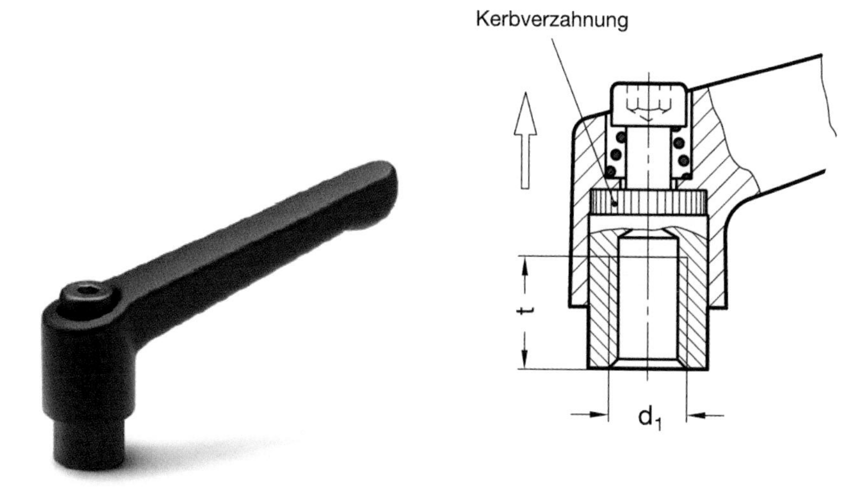

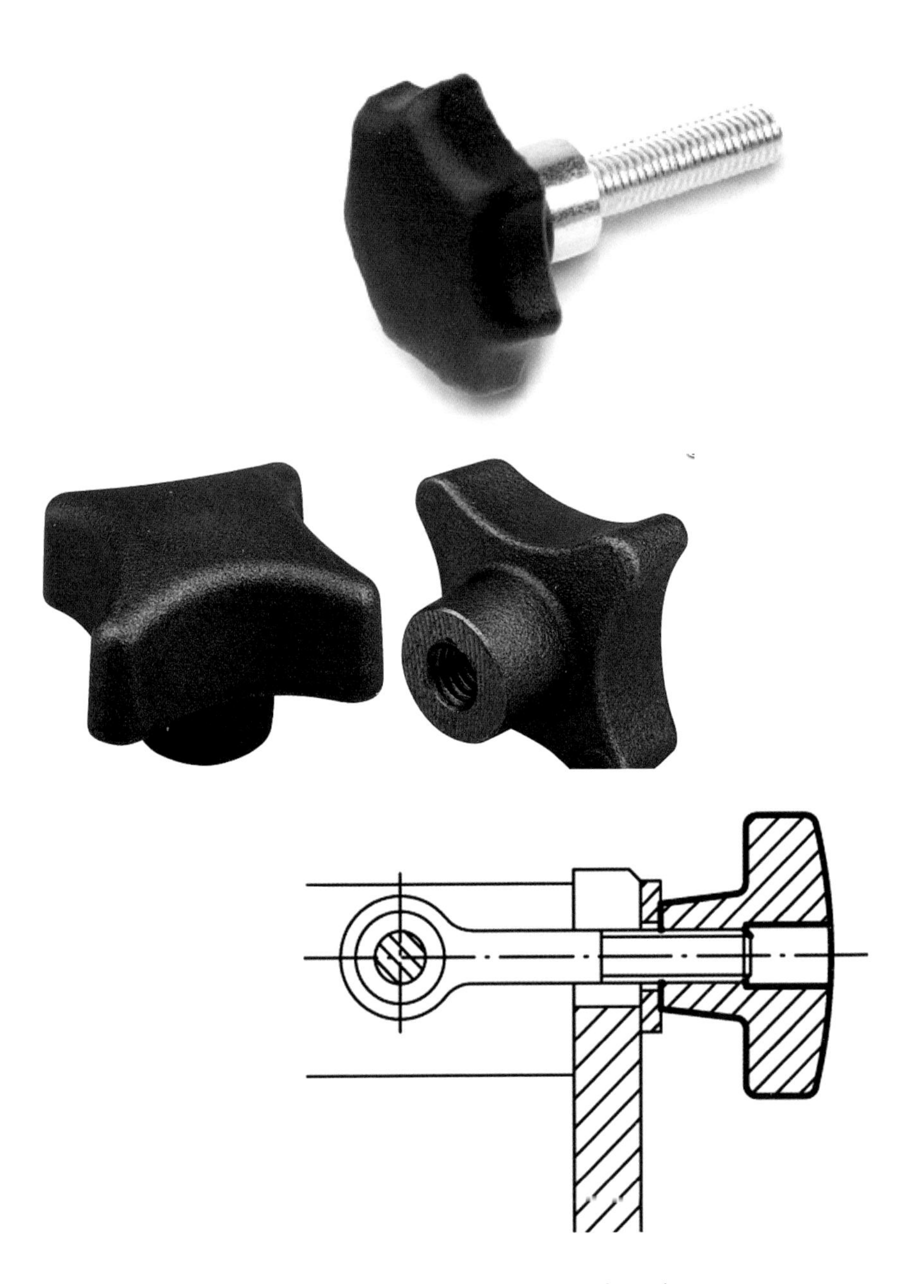

Figure 8-7: Threaded, tool-less usable fastenings.

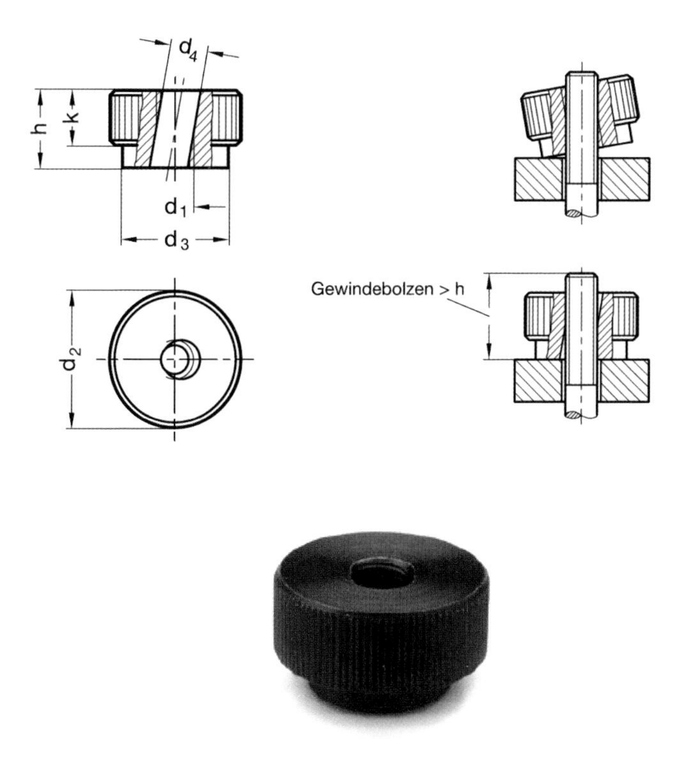

Figure 8-8: Snaplock nuts.

Snaplock nuts are also an interesting alternative, although they remain on threads. Their function is based either, as in the above example, on partial removal of the thread or on appropriate mechanics. They can be quickly loosened, but you must note when using them that they normally represent loose elements, which can lead to the problems already mentioned. There are, however, versions available that are prepared to be anchored at the machine or part by a thin steel wire.

Cam levers are another fastening method which is not based on threads and matches the demands of quick changeovers.

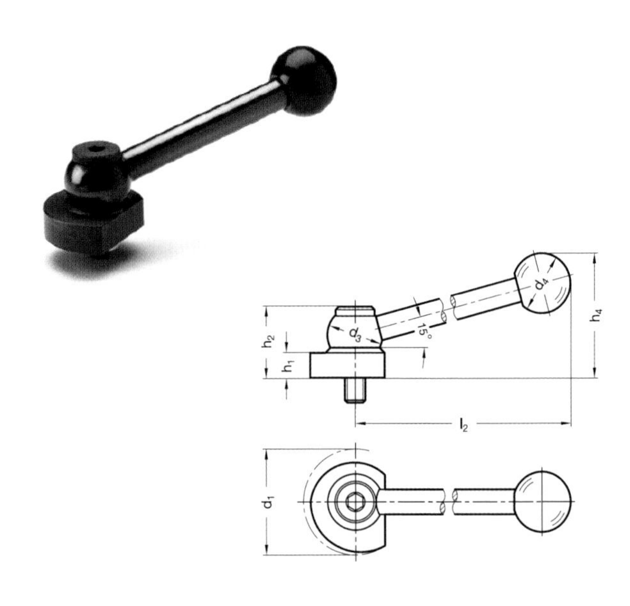

Eccentric cam levers for quick clamped connections

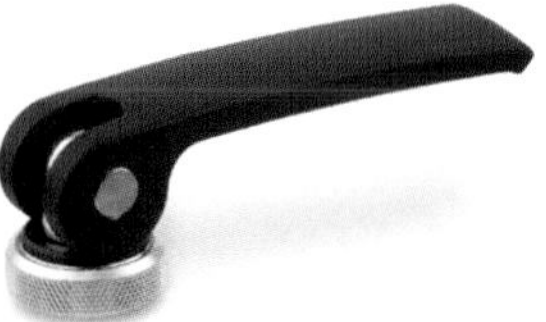

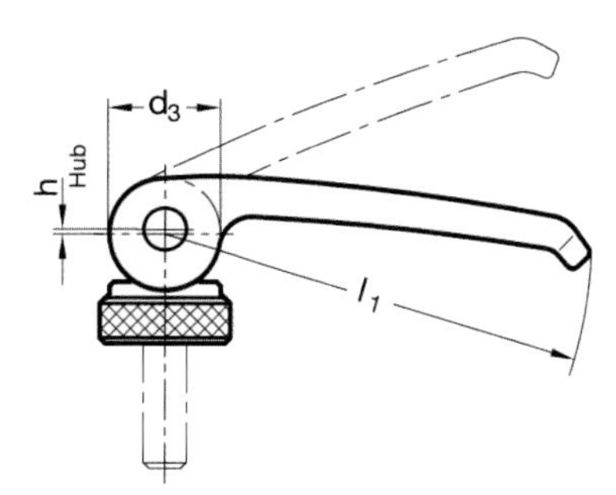

Figure 8-9: Cam levers.

Cam levers are available in various sizes and designs. In combination with eyebolts, they are a very suitable alternative for replacing the screw as a connecting element in many places.

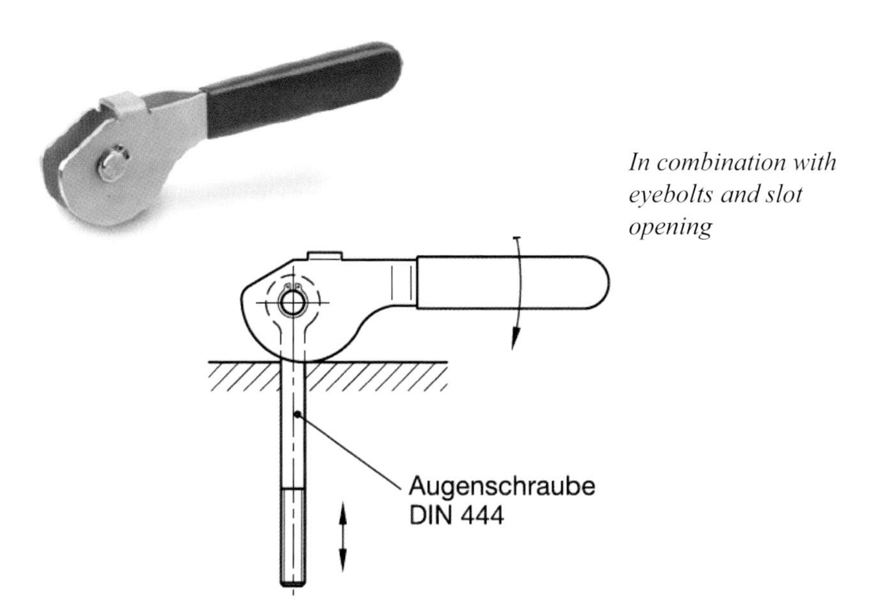

Figure 8-10: Example of an cam lever in use in combination with eyebolt.

The cam lever produces a force-fitting clamped connection, as well as the form-fit represented by the threaded rod. Clamped connections can also be produced with horizontal and vertical clamps. In combination with locking elements attached to the clamp's tip, they can be a useful way to realize form-fits that can quickly be opened and closed.

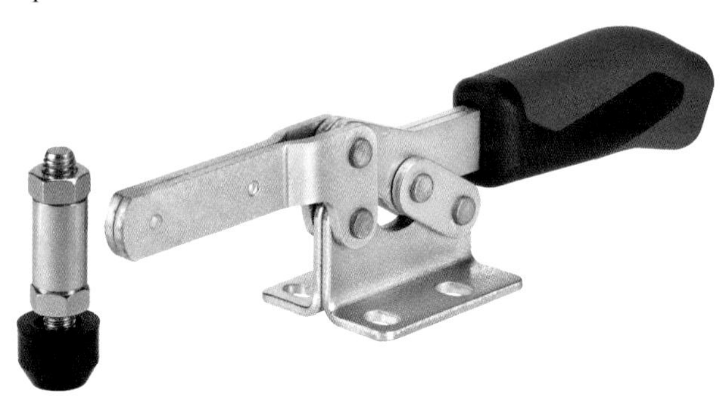

Use horizontal and vertical clamps in combination with inserts and stops.

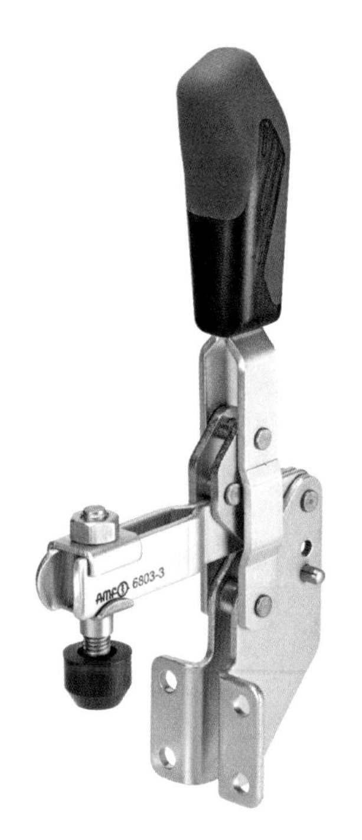

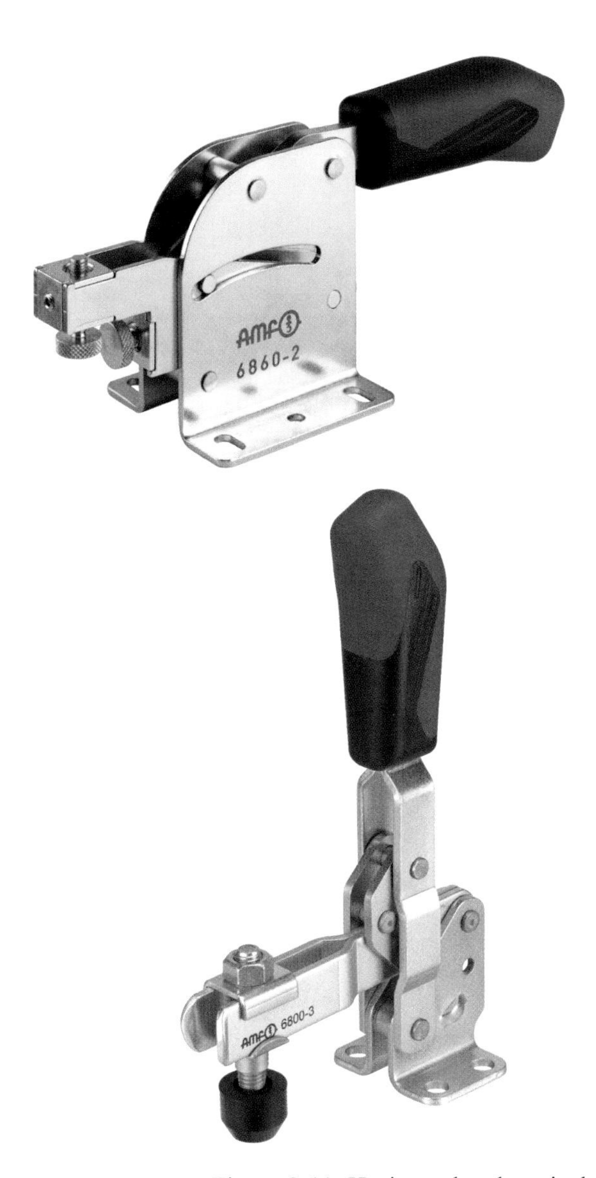

Figure 8-11: Horizontal and vertical clamps.

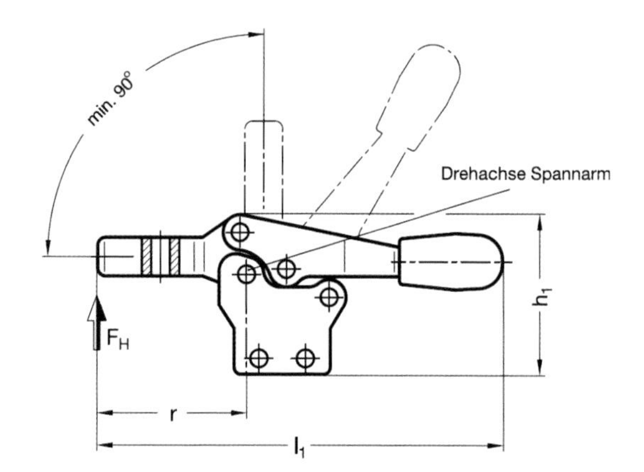

Lever geometry prevents accidental loosening

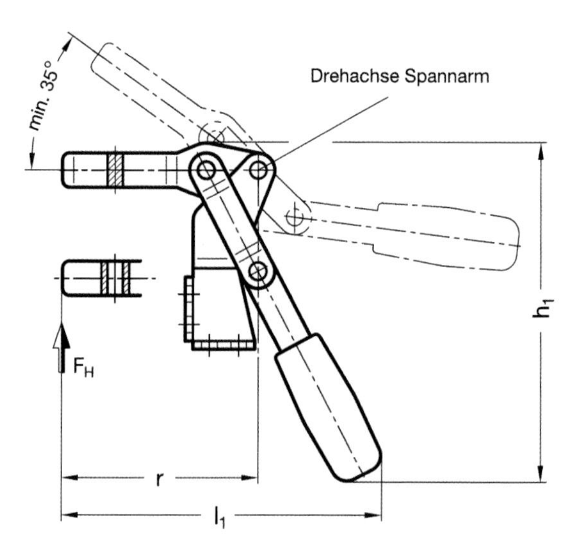

Figure 8-12: Clamp, schematic representation.

Lever geometries are designed in these clamping systems so that automatic opening, caused for example by the process force, cannot occur.

Clamping systems are available both hand-operated, as in the examples above, and power driven as pneumatic ...

Clamps are also available in power-driven versions

Pneumatic

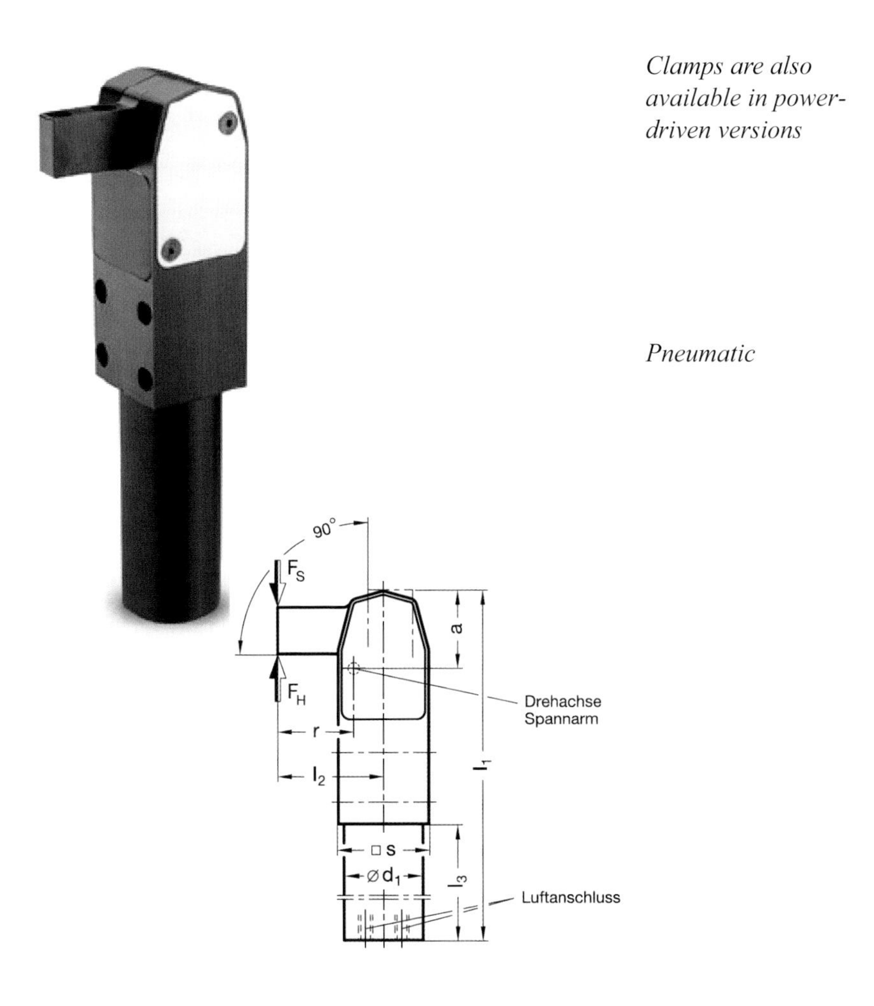

Figure 8-13: Pneumatically driven clamping arm.

or hydraulic versions available.

Hydraulic

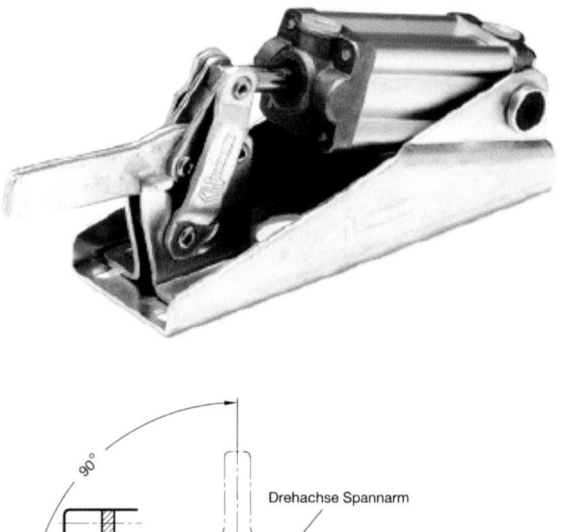

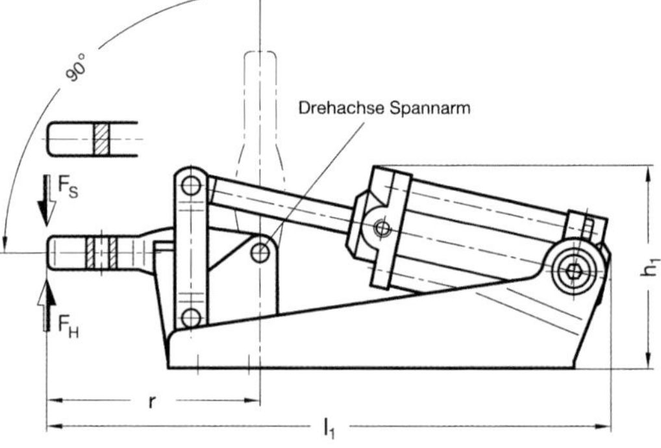

Figure 8-14: Hydraulically driven clamping arm.

Pushrod clamps are similar systems; these are obviously also available in power-driven versions.

Locking and hook clamps are similar systems, which can

support a quick changeover. Systems are available in all shapes and sizes and for almost all force ratios.

Figure 8-15: Pushrod, locking and hook clamps.

Often there can be found reservations among operators as to whether these sort of hooks and clamps are suitable for the situations in their machines. The main concern is that of strength. But that should not be an issue, since strength comes with size, while the advantages come with the functionality.
For the case of lasting concern one could be advised to study the example, of where agriculture, tire extensions are clamped onto tractor wheels with hook clamps. Certainly this is not an environment with small forces.

Figure 8-16: Clamped tire extensions.

Locking connections can be produced with sprung hooks, catches, or in case of lower stresses, with spring-loaded pressure pieces. Both plastic and metal versions are available.

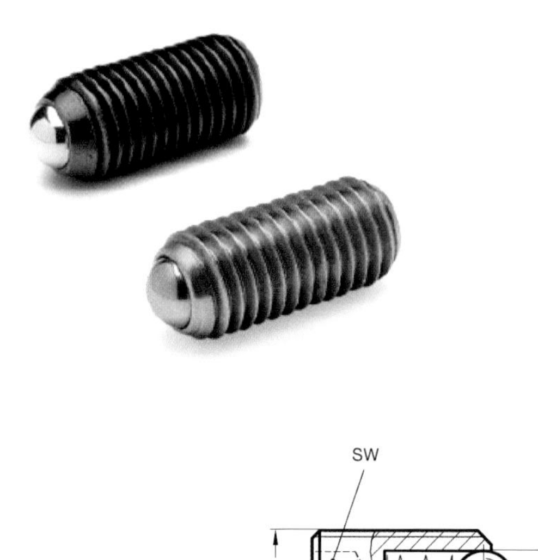

Figure 8-17: Spring-loaded pressure pieces.

8.4 Hose and cable connections

What has been said with regard to the machine elements applies as well in regard to hose and cable connections: Quick loosening and tightening the connection during set-up must be possible.

Interfaces need to be standardized, connection points reduced.

Flow and return pipes need to be combined as far as possible (interface). Zero-tool snaplock connections (GK/Storz/bayonet/combi-plug) are the connection of choice. The industry offers a wide range of systems that can be used off the shelf for the various needs.

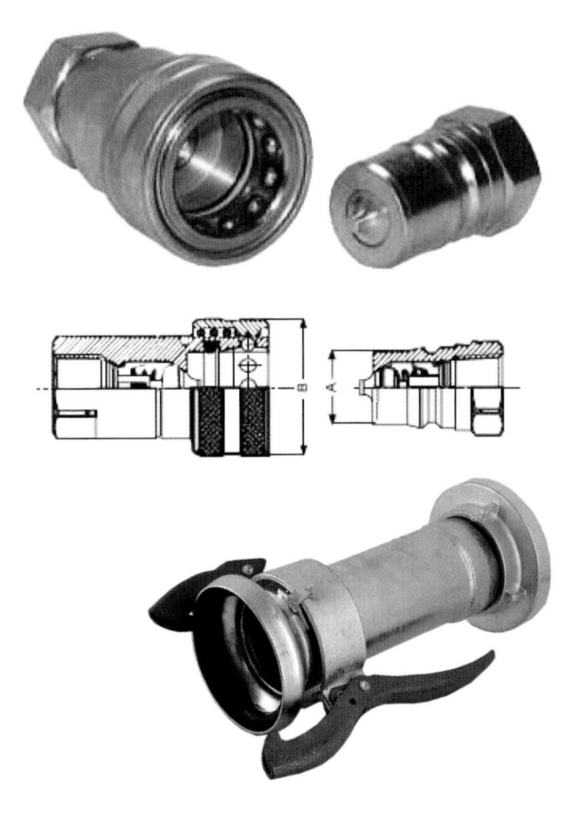

Figure 8-18: Snaplock couplings for hoses and mechanical elements.

9. POSITIONING, SETTING AND ADJUSTMENT

During changeovers, parts, components, and units are exchanged and need to be adjusted in order to be brought into position.
Frequently it can be observed that up to 50% of set-up time is used for these adjustments. The potential for savings is accordingly high.

9.1 Positioning vs. adjusting

In principle:

Positioning takes priority over adjusting!

Adjusting means that the unit or component is first brought approximately into position in order to then be moved successively to the correct position. In this way the accuracy is in the set-up, linked with corresponding time and quality variance. Positioning means that the unit or component is moved directly to the defined position. This will often be realized with the aid of geometry, such as lugs or stops provided in the process. The accuracy is then no longer in the set-up, but realized in the machine itself. Therefore, corresponding time and quality variance are minimized.

Avoid adjustments!

Frequently, the positioning or adjusting of the correct position is linked with the fastening. This is the case, for example, if a component is anchored using fitting bolts. This is very often associated with rough placement and then searching for the precise position. In this case, a functional separation of fastening and positioning is usually beneficial.

Adjustments are sources of variance.

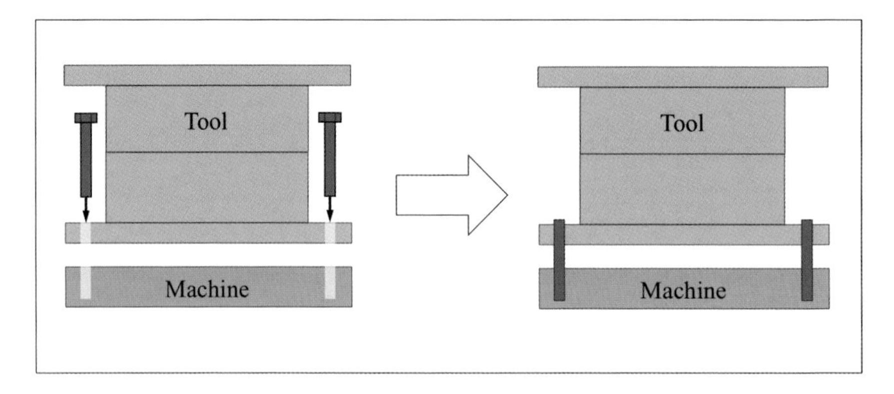

Figure 9-1: Positioning and fastening.

A special case in this regard is a construction where combination of the degrees of freedom of a number of components or units must be aligned.

Accuracy as part of the system, not the set-up

This is the case when units must be both, in a given alignment with each other, and in a corresponding alignment to the process or machine. Significant time savings can be gained if the positioning of the units is fixed relative to each other before starting to tear down. The smaller components are aggregated to a larger component, which can then be positioned more easily in the process.

In the example in Figure 9-2 this can be achieved by the introduction of a filler with the corresponding negative marks of the units, similar to a bite mark.

Figure 9-2: Elements in position relative to each other.

9.2 Adjusting — the right way

Mark product areas to be clearly visible.

If despite all the efforts unambiguous positioning cannot be achieved, adjustments will have to be made. The type and method with which the corresponding settings are realized must be rethought.
In general, it is simpler to move to an area than a given point. Also, defined and marked points are quicker and easier to set than if they are merely marked in the documentation with their settings or coordinates. Therefore, if adjustments are to be made, the corresponding markings should also be made.

Figure 9-3: Areas instead of values.

Areas instead of points

Another subject of question which is interesting here concerns the setting accuracy issue. Working more precisely than necessary does not add value! There can be significant time savings realized by identifying the right accuracy level and adjusting to this, instead of making adjustments to the most "exact" level. Basically, the question is that of the "good enough state".

Remove unnecessary marking.

Also, care must be taken that markings are clearly recognizable and interpretable unambiguously. It needs to be ensured that there are only the necessary markings, that these markings are clearly identifiable and that as few points as possible are needed. It can be helpful in this pursuit to make use of machine properties. Combination of variable and constant positions can offer a range of possibilities.

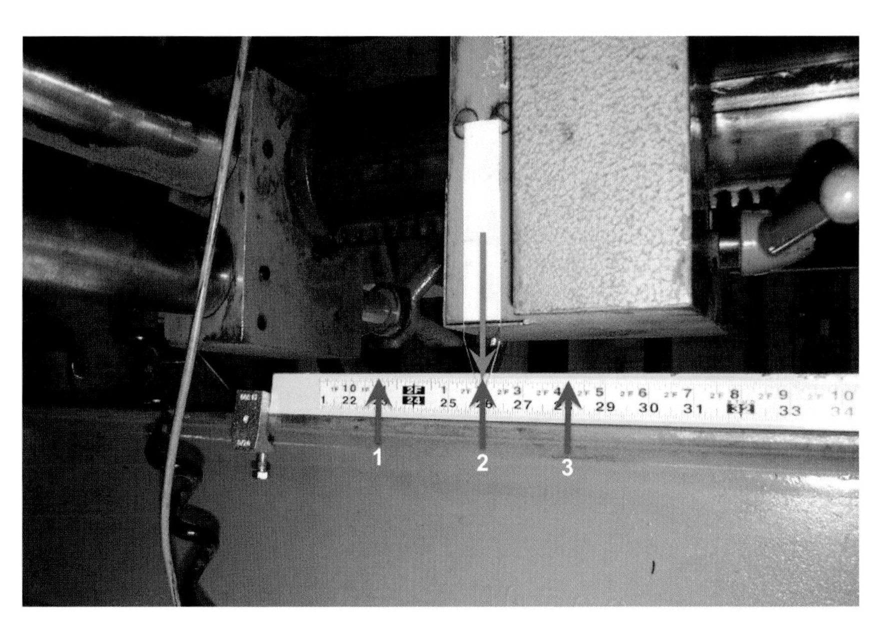

Figure 9-4: Make markings clearly visible.

Use automatic routing to avoid markings.

In doing so, the necessary marking and positioning points could be reduced significantly in a machining center, by including the automatic travel range of the basis of the clamping jaws. It provides virtually <u>any</u> position between minimum and maximum and not only the two, and therefore makes it possible to reduce the actually needed positions improving the situation significantly.

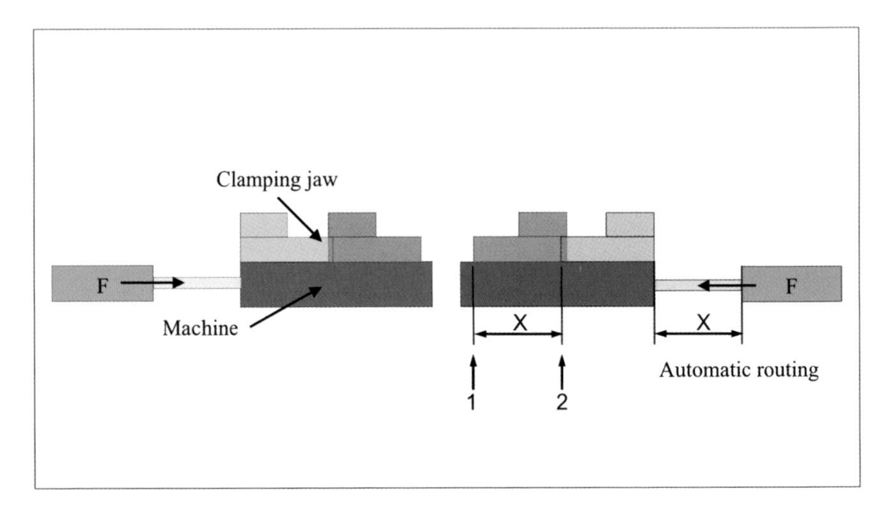

Figure 9-5: Using automatic routing.

Set-ups could be eliminated completely and the security of implementation for the remaining steps was improved significantly, because a confusion of positions far from each other is more improbable than is the case with distances in the millimetric range.

Positioning is by far the more beneficial solution. Where settings are made instead, changes can occur during the process, which can again lead to defective products. In this context, adjustments should be avoided, because they can be made incorrectly.

9.3 Control

Part of a changeover usually is the control of the finished product, to verify the correctness of the settings and positionings made.

Check if possible on the basis of templates.

You should also resort to working with templates if possible for more efficient operations.
If the check is made in a form in which values are determined by weighing, measuring, etc., then it should also be possible to make and adjust settings in the process directly in the corresponding dimensions.

Avoid recalculations; produce tables.

If this is impossible and there is a need for recalculation, tables should be prepared, from which it is clear what effect the change of a setting has in each case on the corresponding target variable in each unit, in order to avoid trial-and-error working methods. Also the probability of mistakes in calculation is reduced greatly if there are corresponding tables arranged.

10. ORGANIZATION

In the previous chapters, theories and methods were worked out to optimize or redesign components and units and their fastenings for a quick set-up. Now organizational aspects of the sequences of the set-up need to be examined. First, the organization of the individual work steps of the set-up are discussed and then the organizational aspects of the working environment will be considered.

Poor organization results in slow set-up processes.

Analyses of changeovers show that a significant portion of set-up time is attributable to organizational reasons. It is not unusual for up to 25% of the time to be used for searching and transporting, as well as on waiting for lifting gears like fork lift trucks and ancillary staff. This proportion can be reduced by appropriate organizational measures, usually without large investments.
It also becomes clear that a significant proportion of work that can be done externally is not — it is done internally and therefore incurs significant down time. This also applies to improving organizational paths.

10.1 Organization of work steps

When analyzing the existing changeover, the individual tasks and work steps are gradually worked out, as described in Chapter 6. Every step is thereby treated on its own merits, without considering further dependencies. However, the individual tasks do not exist unto themselves, but only in conjunction with all the others. This internal organization of work steps must be given consideration, since significant improvement potential can be hidden there.

After using the methods in Chapters 7, 8, and 9 to work out the steps, elimination and simplification, it is now sensible to reconsider a re-ordering of the individual tasks to be carried out.

10.1.1 Externalizing tasks

The re-organization of the individual tasks of the changeover in order to reduce the set-up time always follows this scheme:

Perform external tasks externally; refer also to Chapter 5.

- Avoiding steps completely
- Simplifying the unavoidable
- Performing steps externally as far as possible

The objective is to perform all tasks as quickly as possible and, in any case, outside the machine runtime.
The scheme is followed as long as the time needed for steps to be performed externally is less than the time needed for a product cycle. Once a situation of single-piece flow is achieved, the last step of the scheme is no longer applicable, since it doesn't makes sense to externalize further. What, of course, does make sense in this situation, is to further reduce the time needed for the external steps.

The general principle can be illustrated on a changeover of a production unit.

Once again, the need for a detailed analysis is clear: to gain real advantage, the process must be decomposed into it's smallest operational units.

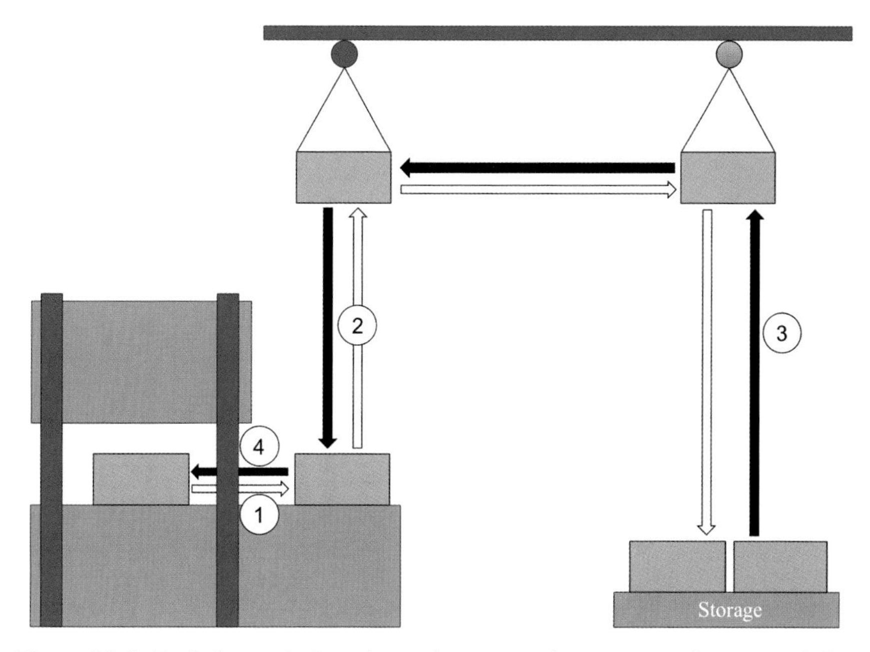

Figure 10-1: Tool change before shortening set-up times: external steps carried out internally.

Analyze work step division: Can steps be performed externally?

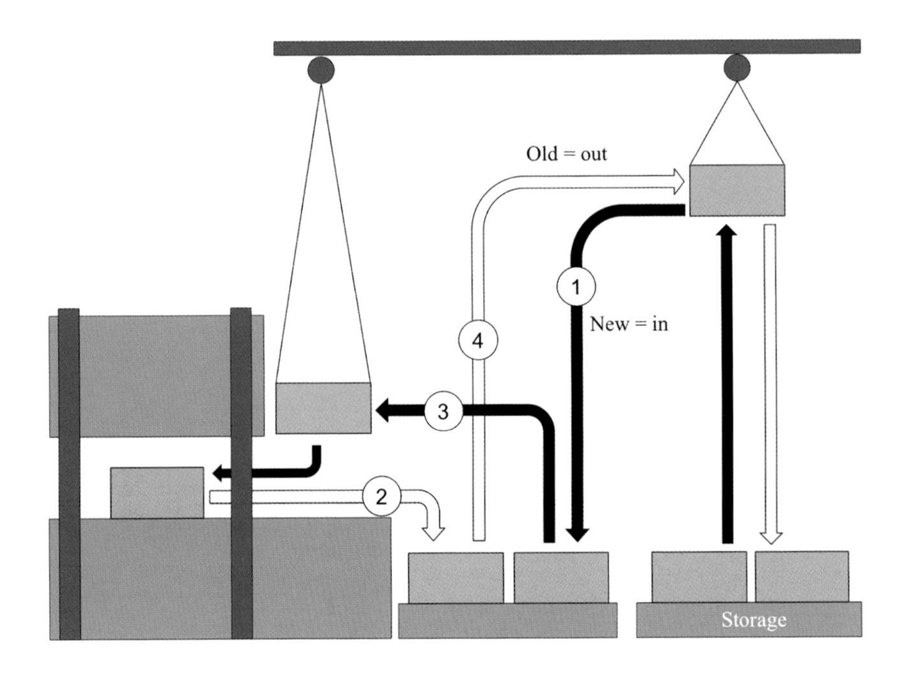

Figure 10-2: Tool change after shortening set-up times: External steps carried out externally.

Installing and dismantling the tool are activities which always take place internally. Accordingly, time can be saved, mainly by eliminating set-ups through product standardizations and by accelerating the fastening of the needed production units and tools. Collection, supply, and removal are not activities for which the process must stop, and accordingly they must be performed externally. In addition, process steps such as heating of tools and carrying out settings, can often be done externally.
To get to the core, it is important to sub-divide the individual steps far down enough, in some cases to the activity level.

10.1.2 Parallel work steps

Performing steps parallel

Individual work steps need to be planned such that, as far as possible, an ideal process is the result. In this regard we can speak of a choreography.
Frequently, all work steps are performed in series behind each other. In this case the end of one step is also the start of the next step. Work steps can be categorized into three different groups:

- Steps which must be performed in their entirety by employees
- Steps which proceed completely alone
- Steps which run partially alone and only require intervention by employees at certain moments

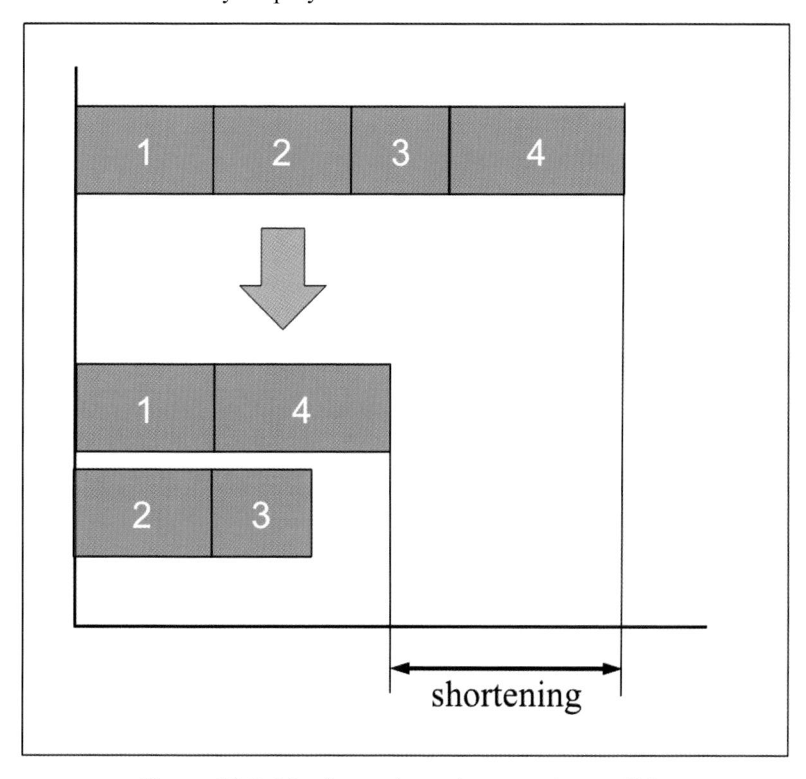

Figure 10-3: Nesting and running steps in parallel.

In order to obtain an efficient total process that is as short as possible, the work steps should be performed in such an order that automatic runtimes in individual steps can be used to perform other steps that need the intervention of employees. This means that work steps have to be nested and steps are run in parallel.

Pit stop as an example of parallel working

A well-known example of a corresponding changeover based on parallelism, among other things, is the pit stop in motor racing. Everything happens in parallel whereby in Formula 1 the vehicle is fuelled and the tires changed within 7–10 seconds, with even cleaning of the driver's visor included in the time.
A similar, from the process outwards, and even more interesting type of pit stop is seen in the USA's NASCAR series. Somewhat different rules apply here and only a few people are available for it. This places even more stringent demands on the choreography of the motion sequences.
To enable parallel execution of the individual steps, it may be necessary to provide additional staff. The pit stop example makes this clear in a very illustrative way. When deciding whether this is profitable, it should be noted that the additional staff can frequently lead to more than just a reduction of the process duration owing to parallelism of the execution of the steps. Frequently, the time needed when more staff are used falls disproportionately, because the need for multiple movements and transport is reduced. In this way, increasing the headcount from 1 to 2 when performing a set-up can lead to a reduction of the time needed by significantly more than 50%.

Additional staff can disproportionately shorten processing times.

So-called Gantt charts can be a useful tool for analyses and redesign of the processes. The individual work steps are depicted as bars, the length of which varies with time and is organized into a corresponding sequence.

Organizing work steps in parallel can be trained in almost all daily routine processes, for example, while brewing coffee, preparing a meal, refueling a car or even changing nappies.

Parallel working in everyday processes

Refueling can be used as a nice example for study as well, to see how big advantages can result from a very small solution: take the little locking clamp that enables the automatic refueling after you started the process. It frees up a significant amount of operator time, which can be used to perform other work steps in parallel and there fore shorten the total set-up time!

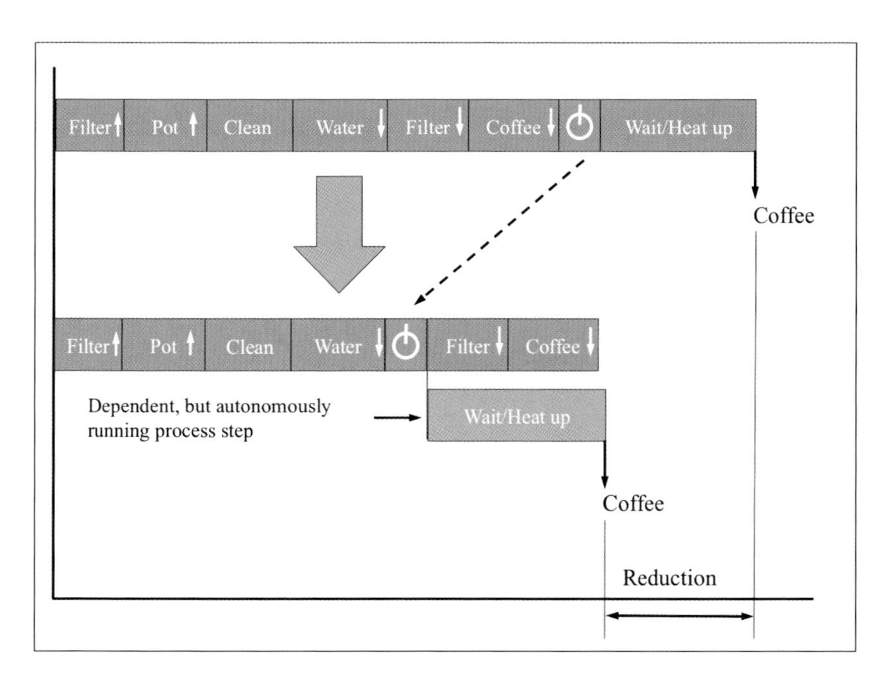

Figure 10-4: Parallel work steps in everyday life: brewing coffee.

10.1.3 Cleaning

Another point for discussion here is the issue of cleaning. Often cleaning is a part of the necessary work during changeovers.
Throughout the process, impurities and dirt must be viewed as critical. They can represent quality-relevant factors which can lead to loss both of product and process. Thus, the work of cleaning needs to be performed meticulously.
In changeovers, however, cleaning is a disturbing factor which should be eliminated. In order to improve the set-up, it is thereby not enough to simply perform the work of cleaning more quickly.

Avoid contamination in the first place.

The optimization sequence needs to be:

1. Prevent contamination.
2. Determine the necessary degree of cleanliness.
3. Search for the simplest method for cleaning.
4. Standardization, avoid "individual" cleaning methods.

Standardize cleaning.

It is often possible to avoid contamination with simple devices, such as bins or deflector plates, or at least make a very obvious reduction. Cleaning should also play a role earlier, when developing processes. Components should be designed so that there are no indentations and corners or gaps in which impurities can gather.

Adjust geometries.

If geometries which may lead to contamination forming exist in the process, these should be questioned. Are all openings definitely necessary? How can openings that are needed only occasionally be closed easily? A plastic stopper to close holes or a piece of hose to fill a T-groove are relatively simple and cost-effective solutions, by means of which cleaning costs can be cut significantly.

The answer to the question of the necessary degree of cleanliness can produce enormous potential savings. Often it does not have to be equally clean everywhere all of the time. Without a doubt, cleanliness is a very important factor in the critical area of the process. But is every area of the process critical all the time?

Separate critical from non-critical areas.

Many roads lead to Rome, and accordingly many methods of cleaning lead to a clean machine. However, there are significant differences in the time taken to achieve the goal.

And, as a final word, "standard" is also the magic word for stable and stably repeatable processes here.
As with all other steps, cleaning should be standardized as well. Together with the principle of avoiding calculations, work can be made easier and variation free by, e.g., having predosed cleaning agents at hand that are the right amount for the standard cleaning bucket. These and other similar solutions can be derived from an integration of techniques like Poka-Yoke, which strives for failure reductions. The very general thought that can be articulated here is, that mistakes and failure are usually a result of a wrong decision or a series of wrong decisions. Therefore, the main question in process improvement needs to be: Where is a necessary decision and how can we avoid this need by introducing clarity?
This helps greatly to reduce variation and make work and life easy.

Poka-Yoke to avoid decisions.

10.1.4 Set-up reduction team

The logic of seeking to shorten individual elements of the changeover and undertake appropriate sortings in the cycle can and should be used not only on the sequence of individual work steps.

At least equally important is dealing with the issue of *who* performs the reorganized steps. Certain work steps in the changeover require knowledge of the machine and of the production process and are critical to some extent. This includes activities that are directly related to preparation for production, hence, setting up and adjusting the machine. Accordingly, this requires an appropriately trained employee, frequently the machine operator, for execution.

Have work steps performed according to skill level.

However, not all work steps belong in such a critical category. Transport activities and the work steps included for teardown are often substantially less critical and therefore require no specialist knowledge of the machine for these to be performed with satisfactory results. Principally, it is wasteful to occupy highly skilled and therefore expensive and scarce personnel resources on tasks whose demands do not match the employee's capabilities. In particular, in situations where an (apparent) lack of specialists prevails, it is therefore sensible to ask the question “Who does which tasks?” It can become interesting relatively quickly to train special set-up workers or set-up teams, or to allocate individual activities that are less demanding to appropriately skilled personnel.

Take the load off specialists.

10.1.5 Information flows

The final point in this section will discuss the topic of information flows.

Complete administrative tasks externally.

Frequently, a not inconsiderable part of set-up time is used for processing and editing information. This includes administrative tasks such as searching for, producing, and loading programs.

Whenever possible, administrative tasks should be performed externally and remain restricted to what is absolutely necessary. This means separating variable and constant factors from each other and deciding what needs to be communicated. Noticing that there is nothing to note is a pure form of waste. This is at its worst when it occurs in printed papers and on forms. These "non-communications" will then tie up employee capacity again in the ongoing processing sequence.

Realize end-to-end digital task structures.

Data structures should be organized comprehensively in a digital manner and need to be centralized. It is still a widespread practice to print out previously digitized data and then enter it into the production system, digitizing again. Production feedback is then again organized accordingly, which generates a chain of work steps that are alternately analog and digital. These constant conversions represent waste, are slow, and promote the occurrence of errors. With such structures, it will hardly be possible to occupy a leading position in the industry.

Store information centrally

The requirement for centralization is related to the issue of storage location. If data such as production programs strays on all kinds of local data media, chaos is preprogrammed. This includes both USB sticks and filing in machine memories. As soon as the first, however small, version change is made, there is no longer any guarantee of smooth functioning. There is also a high risk that information will fall into the hands of unauthorized third parties.

Filing information and programs in folder structures that have to be explored by hand should also be prevented if possible. It should be possible to call up by entering the program number in a search field or by scanning in a barcode.

If, despite all your efforts, working with a folder structure is unavoidable, the organization of these structures must be done carefully. There must be a logic implemented that covers all processes of the company, and that must be as simple as possible to work with. In relatively stable production schemas, this can also be provided, for example, by storing frequently needed programs "on top" to avoid multiple clicking.

Order in the database as well: First things first!

Decentralized generation of production programs on the machine, as is common in some industries, represents another challenge. This method of working is questionable, because there is usually no central storage and there is a probability of errors due to the programmer's attention being diverted by the various disturbances to which he might be exposed. It speaks for itself that programming also must not result in machine downtime; this work step needs to be performed externally.

Data should always be stored centrally and released for one-time use for performing the machining.

What has been said already about mechanical degrees of freedom also applies, particularly for the information structures area. Fixed places, for example for tools, must be assigned in information and program structures.

Avoid manually entering information.

If information is already stored in production programs, this must be read out and cleared for correctness, in order to avoid errors and save time. The need for repeated entries must be avoided as much as possible.

10.2 Environment organization

10.2.1 Workplace organization

After the internal complexity of the changeover has been reduced by performing the steps described in the earlier chapters and the individual work steps have been organized, the working environment in which the changeover takes place must be organized accordingly.

Order and cleanliness are the foundation for quality.

The lean tool box includes the tool of 5S.

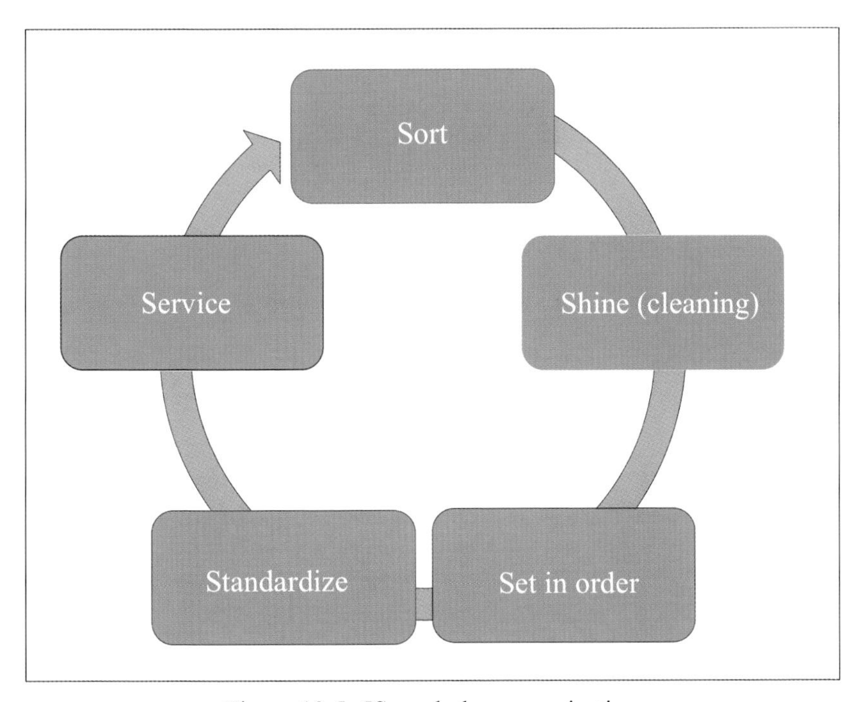

Figure 10-5: 5S workplace organization.

It is a method to design workplaces and their environment so they are safe, clean, and well arranged.

Order and cleanliness are basic conditions for improving work processes. Therefore, it is very common to start production improvement programs with a 5S campaign.

5S stands for the five steps, each beginning with the letter "S":

Perform 5S as PDCA.

- **S**ort out
- **S**hine (clean workplace and work resources)
- **S**et in order (put everything in its proper place)
- **S**tandardize
- **S**elf-discipline so that order and cleanliness are maintained

Cleaning before reorganizing

Often there will be found a similar row of terms, where the second and the third S's above are exchanged. In the author's view this is not correct. I would suggest you first clean things and places, and then put them back in an orderly manner. Having lead many 5S workshops and having worked with companies that are regularly using 5S as a tool, I have never found a better way of doing this then the one suggested above.

The objective of a 5S program is to design workplaces so that the work can proceed without interruptions, searches, and long transport routes, and waiting times are avoided and therefore the work can take place as much as possible without waste. A clean and tidy working environment is the basis for sustainable high-quality work.
Visualizing the standards that have been developed also helps to support sustainability. In this way, deviations can be detected and rectified quickly.
Within a set-up reduction program, a 5S campaign will have a prominent place as well.

10.2.2 Visualization

The need for visualization was emphasized earlier. Both in designing the process environment and in the process itself, markings are a means to display information in a method and manner that is unambiguous and cannot be misinterpreted.

Visualization for quick identification

In the process environment, clearly ordered structures must be realized in which everything movable has a firmly assigned place. Colored markings are the method of choice. Symbols, letters, and numbers represent an extension of the options, with symbols being preferred over letters, because they can be interpreted more quickly. In any case, universality of the logic used should be observed.

Colors and symbols before numbers and numbers before writing

The simplest way to achieve this for tools is with shadow walls, which reproduce the outline of the corresponding device. In this way, it is easy to see which tool is missing. Tools and other aids should be available in adequate quantities for every process. Clearly recognizable markings must be applied accordingly.
Tools and aids for process "A" will then be marked accordingly with a red "A."

Mark storage areas and tools.

Likewise, components and units should be stored in an orderly manner. For this, an appropriate marking system is needed which supports easy finding. In storage locations, columns lettered A to Z and numbered rows can mean significant simplifications in this case.

10.2.3 Storage

In principle, it is desirable to always store process-related parts and tools close to the relevant process in order to achieve short set-up times. This contradicts the practice frequently encountered in companies of storing all parts at a central site, far away from the processes, on the pretext of apparent order and visibility. The transport associated with this practice should obviously be viewed as adding zero value. Ultimately, process organization should be about effectiveness and efficiency, not about beauty.

Avoid central storage.

The organization of the resources and components should be aligned with the frequency of use in each case.

Free areas. Reduce the need for movement.

If components cannot be assigned fixed places, it should be ensured that shelves and tool trolleys always have space available. That way, it is possible to provide the removed component with a storage area immediately and avoid prior handling and moving around to free up the needed space.

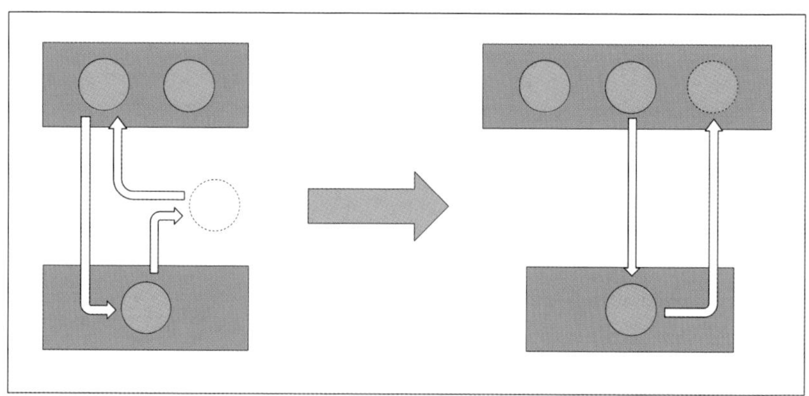

Figure 10-6: Free storage area for movement and transport optimizing.

The objective of set-up reduction programs is not to set-up less, but more often. Only this leads to the necessary flexibility. From this follows:

What is often produced, is set up often!

Store items that are needed frequently close to the place of use.

Therefore, relevant parts should be stored close to the process.
If this is impossible for reasons of space, all parts for the process should be stored so that storage areas are allocated in a priority order. Things that are needed frequently must be stored so that they can be reached easily.

Leave units and tools in the process if possible.

When organizing storage sites for parts to be moved by hand, an ergonomically sensible layout should be ensured. If in doubt, an ergonomically beneficial allocation takes priority over an exact hierarchy. This means that heavy parts are stored at an appropriate height so that moving and accommodating them is simplified. These parts should always be stored as close as possible , or, if possible, inside the actual process.
Frequently, the conditions needed for this can be achieved in a relatively uncomplicated manner.
In the actual case of a concept for a stamping machine, working with a simple ball caster track made it possible to leave the tools inside the process. The stamping tools are simply pushed in and out of the stamp. This eliminates the need for transport to the shelf and the corresponding storage area.

Figure 10-7: Ball caster track.

However, take care: It is not enough to just rearrange the storage of tools.

Caster roller tables and roller tracks as unit stations

The market dynamics mentioned at the beginning makes it necessary to update this arrangement of tools at regular intervals. There should be a process defined to revise the organization on a regular basis to check whether the arrangements are still the optimal ones or whether adjustments are needed.

10.2.4 Transport

As mentioned above, transport must be avoided, because it represents a zero-added value activity or in lean terms: waste. If this is impossible, then appropriate arrangements should be provided such that the most frequently needed items need the least transport, in order to achieve total transport effort that is as low as possible.

Match transport equipment to needs.

To reach this goal of lowest effort, means that steps need to be taken to simplify transport as much as possible. For example, individual parts to be set up can be combined on set-up trolleys. If parts and units are very large and/ or heavy, the use of special set-up trolleys and the use of locally fixed lifting gear can be sensible. Pallets as a means of transport which is also encountered frequently, should be avoided if possible. Components are housed on these in an unspecified position and can easily fall off during transport, be damaged in other ways, remain suspended in free slots, or simply be lost.

Produce set-up trolleys.

Avoid standard pallets.

Another form of transport optimization can often be achieved by using roller and caster tracks, as mentioned above, or other storage options directly in the process, on which the corresponding units and tools can be placed for intermediate storage. This is mainly an option for the tools for A-products, since these are set up accordingly often.

11. COMMUNICATION

A newly installed changeover procedure must be documented and communicated accordingly. If these are the first efforts to reduce set-up times, the communication must encompass the entire company in order to inform other departments of the changes as well. The individual elements of the set-up are often repeated at other machines and should be standardized accordingly. In order to achieve production flexibility in the long term, employees must be capable of rotating, in order to master a number of processes or to be instructed quickly in new processes. In this context, reducing complexity by introducing standardized elements is a major contribution. Standardization is therefore a value far from being overestimated, not only inside the process but also beyond process boundaries.

Communicate across the entire company.

11.1 Documentation

Processes must be controlled in order to be successful with flexible production systems. All employees entrusted with changeovers therefore must have the same information concerning set-up procedures and use this to perform the changeovers in a standardized manner.

Make information accessible.

For standardizing the procedures and sequences of the individual work elements in a changeover, it is useful to apply a workbook in which the sequence of the work elements is depicted step by step. For this purpose, boards are available in which each step can receive a separate table, to avoid confusion and ambiguities. In the actual description, you should resort to images and visualizations as far as possible. A picture says more than a thousand words and descriptions can often be misunderstood. Also, visual elements can be interpreted substantially more quickly than text.

Avoid "private" information and "secret" settings.

Figure 11-1: Board system for step-by-step depiction.

11.2 Before and after

Document before and after visibly.

In order for the successes that have been achieved by the efforts to shorten set-up times and other optimization programs to gain permanent and sustained appreciation among employees, it is immensely important to compile good before-and-after documentation.

Corresponding pictures and relevant process parameters should be placed on site so as to be clearly visible. Where a process originates, its current position, and where it is to be moved must be clearly and unambiguously documented. After a certain time the workforce can become accustomed to the new situation and an undertone enters echoing the past, which results in the motto “it wasn't so bad before.” This then becomes the starting point for further improvement projects, which can hinder starting up, since one sees the efforts to come but may underestimate the gains.

This also means that changeovers should be documented constantly. Each set-up is to be time-recorded and the data needs to be analyzed. In this way, the topic remains current and can create an objective basis for additional improvements, because each change recommendation and every idea can be checked immediately for its effects. Also, an unwanted deviation can be recognized at an early stage in this way and suitable measures taken.

12. FINANCIAL ASPECTS

As with all efforts at optimization, the objective of a changeover reduction is to improve the situation.
It is relatively easy to illustrate whether the changeover in the new situation needs less time than before. It becomes more difficult if the question has to be answered, whether this gain in time also represents an improvement from a commercial standpoint.
Finally, the crucial point with all investments is always, of course, the answer to the question of when the expenditure is returned.

Justify disbursements.

The actual problem here is not in the calculation. Every company has to decide for itself on appropriate methods for assessing investment issues. The problem is the question of the basis for calculating the earnings side. Depending on the standpoint, this can vary widely. It happens that recommendations for improvement in the company are rejected because it appears impossible to calculate an outlay on the basis of workforce hours or machine hourly rates.

Choose the correct calculation basis.

It is essential for this to create clarity concerning the situation. There are basically two result areas to distinguish — internal and external.

12.1 Internal

Set-up reduction campaigns will make a contribution to reducing inventory if used in accordance with the logic of synchronized production systems. This is a direct, active contribution to reducing capital circulation time, which leads to a reduction in working capital financing requirements.
Accordingly, on the credit side, at least the weighted average cost of capital (WACC) for the inventory that doesn't have to be pre-financed, should be appropriated.

WACC as calculation basis, etc.

Apart from this, a few somewhat more tiresomely quantifiable issues must also be regarded:

- Low inventory levels mean less storage and accordingly less storage area is needed. If a warehouse expansion considered as necessary can be prevented, this can generate significant savings.
- Transport and administration are reduced by reducing inventory requirements.
- Transport routes are reduced through smaller storage areas and the extent to which transport resources and staff are tied up accordingly.
- Less must be produced on predictions and forecasts, and therefore the probability to have to write-off for excessive storage and old inventory will fall accordingly.
- Irrational capacity in the production system can be activated, which leads to increased productivity figures.
- Accelerating changeovers means more frequent set-ups, which results in the staff getting more versed in regard to performing the changeovers. This reduces variations in the changeovers and therefore in the production schedule, which again leads to higher productivity.
- Should there be a real bottleneck in capacity, the savings of the campaign results equal the extra profits made due to the increased capacity.
- If in the same case, a new investment is postponed or totally obsolete, the corresponding capital costs are to be appropriated.

..... and weight qualitative factors

Costs of machine downtime is another measure that is frequently used to examine changeover reductions. This can deliver impressive numbers which helps to get a change going. It is, however, most often not an appropriate figure, since it only holds in case of real bottleneck issues. As long as there is excess capacity, this is obviously not an interesting issue.

It becomes clear that the effect of set-up reduction campaigns leads to a complex credit contribution in the internal consideration.

12.2 External

Determine strategic factors.

However, being able to perform quick changeovers is relevant not only from an internal point of view but also in external regards. Some points which should be quoted here in connection with the added production flexibility gained are:

Failure to implement can mean market losses.

- ➯ The capability to realize more complex customer value propositions
- ➯ Greater delivery flexibility; JIT and JIS delivery
- ➯ Improved ability to respond to changes in customer requirements
- ➯ Opportunity to increase market shares through improved competitive position
- ➯ Entering new markets by increasing production logistic capabilities (e.g., ease of audit for large industrial deliveries)
- ➯ Better capital requirement structure

Finally, these external factors will turn out to be the real reasons for having short set-up times. As the market dynamics grow, the need for ever higher levels of reactivity will continue to accelerate, and it will finally become a question of survival, not only of success.

Acceptable levels of set-up time in new machines?

This lays a very special demand on companies that are at the point to invest in a new machine. How long are the set-up times in this machine? And for how many years will this be competitive, if it is at all?

Careful consideration needs to be taken in these questions, especially due to the exponential growth rates mentioned above.

13. IMPLEMENTATION EXAMPLE

Finally, the implementation of the methodology is depicted with some examples and the results achieved are shown.
For reasons of improved representation, models will also be used extensively.

Throughout the book, several theoretical frames and tools have been developed to understand the nature of a changeover and shorten set-up times. It should be noted, that a changeover will usually be a combination of factors, elements, and categories that can be improved by the use of these tools.

Changeover as unique combination of structures

The regular questions and discussions about changeover reduction that follow a format of "*who has an example for shortening set-up time on a xyz-machine,*" which can be found in forums frequently, are missing the point.

Just as demonstrated at the example of forces in relation to their directions, changeovers need to be examined thoroughly and improved step by step. Determine the factors that are there, analyze thoroughly what is the actual cause for that factor, and find a solution for that factor is the preferred way of working. Usually there won't be too many answers found at the aggregation level of a whole machine.

13.1 Press

In an industrial press for stamping steel bushings, the set-up time for changing the pressing tools was analyzed. The 40-mm-long bushings are pressed in predrilled plastic plates, 30 mm in thickness. The end products differ in the number and coordinates of the bushings.

13.1.1 Functional analysis

The tool for stamping the bushings is a massive steel plate. The bushings are fixed to their coordinates by pins that are pressed into the plate and which protruded 10 mm out of the plate. The functional analysis of tool and product showed an interfacing problem: The pins by which the bushings were positioned determined the outer geometry of the tool and made it impossible to produce several products on the same tool. A functional and process force analysis revealed that the bushings had to be held in their position and that the process force was diverted cleanly along the longitudinal axis of the bushing. The tool design was then changed so that the pins were replaced by borings.

Refer also to Section 7.2.

The bushing is no longer held on its coordinates by the pin from the inside, but by the baseplate from the outside. In this way, it was possible to use the tools with a number of products. It was possible to combine the 5 most produced products on a single tool, meaning that a significant number of set-ups could be eliminated completely.

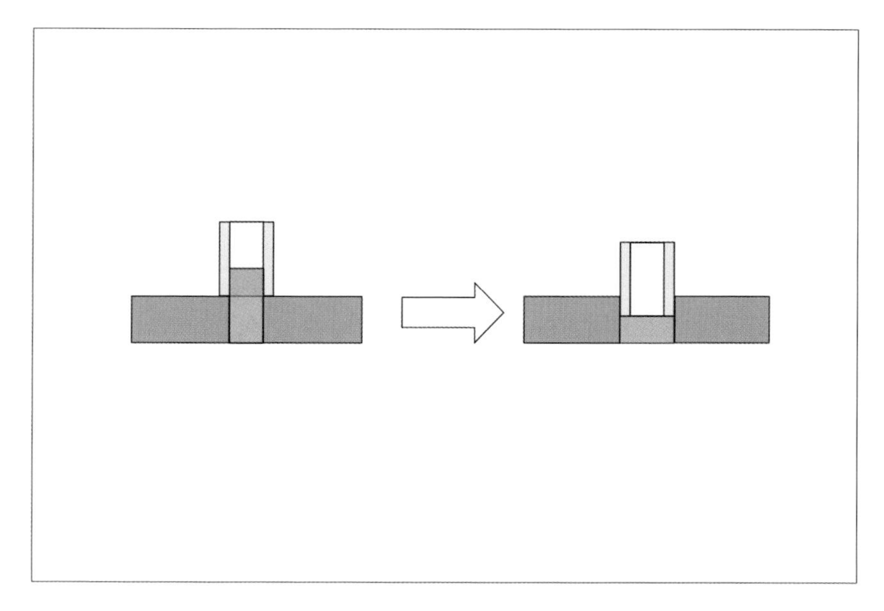

Figure 13-1: Result of functional and process forces analysis.

13.1.2 Process forces analysis tool

The process forces analysis revealed that the main process force was diverted cleanly along the longitudinal axis of the product. Accordingly, less force is required to hold the bushing at its position. It also became clear that the process force was diverted by the material directly below the bushing in the machine. The tool material was changed accordingly to match its functional requirements. The massive steel plate was replaced by an HDPE plastic plate, in which steel inserts were placed at the appropriate coordinates, by which the shells were held and stamped.

Refer also to Figure 9-1.

The plastic is strong enough to hold the steel inserts in position such that they divert the process forces.

It was possible to reduce the weight of the tool by 75%.

13.1.3 Tool fastening

In the original design, the tool was connected to the machine by means of screws.
The tool was inserted at the approximate position in the machine, and after the position was found, secured with two screws.
The process forces analysis did not reveal any forces directed upwards. Accordingly, it was unnecessary to suppress the corresponding degree of freedom.
The screws were replaced by fitting bolts, which were placed in the machine's tool housing. The tool is now merely inserted into the machine or withdrawn. The fitting bolts prevent movement and rotation. The step can be executed without tools and without loose fitting elements.

13.1.4 Summary of the results

Changeover time reduced from 900 to 10 seconds!

Tools are no longer required. There are no longer any loose fastening elements. Internal assembly in the tool (pin changes) was eliminated. The tool weight was reduced by 75%.

Process integrated in one-piece flow line

In summary, the set-up process at the press was reduced from 15 minutes to 10 seconds. The press could therefore be integrated as a process into a one-piece flow line, eliminating all side effects of batching (transport, administration, storage, etc.).

13.2 Sheet bending line

In a sheet bending line, the different elements of the changeover were analyzed and improved. Various different rails were exchanged and very heavy stamping units had to be exchanged.

13.2.1 Rails

To change various rails, bolts had to be loosened using an Allen key. Eight bolts had to be loosened on each rail. The rails were at a position in the machine which could only be reached with difficulty.

Change without use of tools

The process forces analysis revealed that the process force had to be diverted in its vectors (degrees of freedom). The different rails were pressed upwards and towards the machine housing.

The rail fastenings were changed to match the functionality of a bayonet closing. Millings were made in the rails, by means of which they were pushed onto guide pins that were placed in the machine. The guide pins were fitted in borings on the machine's tool housing, which were used earlier to house the bolts.

Loose elements completely avoided

At one point, a clamped connection was provided by means of a cam lever.

Changing the rails now means only loosening the connection by operating the cam lever, moving the rails out, pushing new rails in, and remaking the connection.

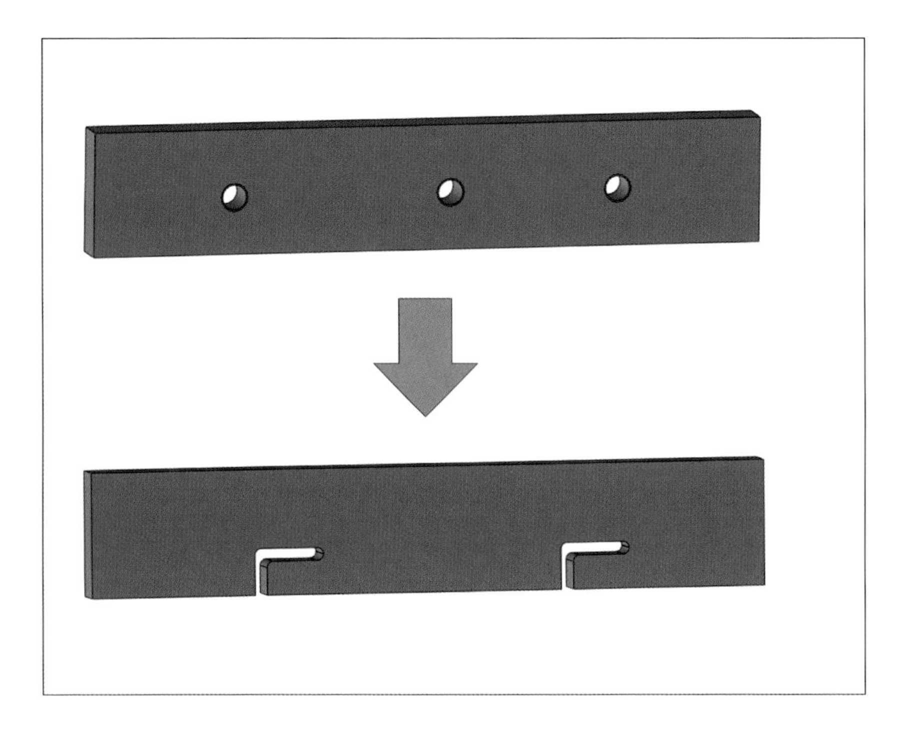

Figure 13-2: Bolted connection replaced by bayonet connection.

Reduction: 35 minutes down to 1 minute

The task is done without loose fastening elements and without tools. It was possible to reduce the entire rail change time from 35 minutes to 70 seconds and the ergonomic situation was substantially improved.

13.2.2 Stamping unit station

Reduction: 1 Person

During the changeover, a number of stamping units had to be exchanged. To do this, the units were removed from the machine and stored in a high rack storage nearby. The exchange sets were also taken from there. Due to the great weight of the stamping units and the somewhat inaccessible situation, this task had to be done by 2 people.

A “station” based on caster rollers for stamping units was created at the machine, directly on the housing. This means that the stamping units can now remain in the machine. In the new set-up process, the stamping units only have to be pulled out in each case, moved to a free area on the caster, and a new unit pushed in.

13.2.3 Organization

The finished product is bulky and is packaged in steel-framed pallets for in-house storage and transport. The empty steel-framed pallets are stored outside the building. By shortening the changeover time and using the corresponding opportunity to work with smaller batch sizes, it was possible to substantially reduce the number of steel-framed pallets to be stocked. Storage space was freed up by reducing the number of production batches. The storage point for empty pallets was moved to these free storage areas, directly next to the full ones. This halved the need for supply runs around the production process, because the previously empty return journey from the storage place is now used to bring a new empty pallet.

Reduction: 50% storage space, less transport.

Because the total storage area now needed is less, the storage can be closer to the process, which also leads to a reduction of movement and journey times.

13.2.4 Summary of the results

The total set-up time is reduced by the described and other solutions from over 72 to under 8 minutes.

Reduction: 72 minutes down to 8 minutes.

13.3 Roll-forming machine

On a roll-forming machine, the set-up consists of changing different rolling, profiling, stamping, and cutting units.

13.3.1 Rolling and stamping units

The storage rack for the units, directly next to the station, was completely full.
Units were exchanged using a forklift truck and a lifting gear consisting of chains. The lifting gear was mounted on the forklift, the unit then suspended on it and moved away from the process. Two people were needed.
The set-up consisted of these steps:

- Fetch a new unit from the rack.
- Collect and mount the lifting gear.
- Remove the old unit from the machine.
- Dump the old unit.
- Take up the new unit.
- Mount the new unit in the machine.
- Remove the lifting gear and put it away.
- Take up the old unit.
- Put the old unit in the rack.

The storage situation was expanded by defining a free space.
The transport function was integrated by fitting skids in the units. Lifting the machine created an opportunity fpr underpinning.

Refer to Figure 10-6.

The changes made it possible to perform the set-up with a walkie stacker and without lifting gear. The free space eliminates cumbersome double handling.
The new set-up can be performed by one person and consists of these steps:

- Remove the old unit from the machine and place

it in the rack.

- Collect the new unit from the rack and mount it in the machine.

13.3.2 Summary of the results

- Expensive forklift truck no longer required
- Set-up time shortened from 22 minutes to 3 minutes
- No loose lifting device (searching for it, collecting it to bring to the process, putting it away, risk of accident)
- One person can perform the set-up.

Reduction: 1 Person

The procedure is consistent with the functional and interface analyses mentioned earlier: integrate functions, avoid isolated units.

13.3.3 Cassette principle with cutting unit

The cutting unit consisted of a hydraulic unit and various cutting tools.

The different cutting tools were bolted to the hydraulic station. In each case, the station had to be matched to the various dimensions (variable height and width) of the tools.

Figure 13-3: Cassette principle, form-fit, horizontal clamp.

By standardizing the external geometry of the various cutting tools to the dimensions of the largest unit, following the cassette principle, it was possible to eliminate matching tasks on the station itself completely. The fastenings of the units in the station were changed from bolted connections to clamped connections made with a horizontal clamp. The connection with the hydraulic unit was changed from a bolted connection to a form-fit push-in connection. This was possible, since the process force analyses showed clearly, that there are only force vectors along the longitudinal axis working.

13.3.4 Summary of the results

It was possible to reduce the set-up time at the station from 28 minutes to 15 seconds. No tools are required for the work at all. All loose fixing elements were eliminated.

1680 seconds down to 15 seconds!

Reduction: 100%!!

Bibliography

In the course of the book, further literature has been mentioned on a number of occasions; a complete list with title, author, and ISB number follows here:

What Is This Thing Called Theory of Constraints, and How Should It Be Implemented?, E.Goldratt, ISBN 978-0884271666

Factory Physics, Spearman and Hopp, ISBN 978-0071232463

Quick Response Manufacturing: A Companywide Approach to Reducing Lead Times, Rajan Suri, ISBN 978-1563272011

My Work and Life, H.Ford, S.Crowther, ISBN 978-1162577944

Learning to See, M.Rother, ISBN 978-3980952118

The Toyota Way, J. Liker, ISBN 978-3898791885

Einstieg ins Systems Engineering, R. Züst, ISBN 978-3857437212

Dr. Russell Ackoff on systems thinking, http://www.youtube.com/watch?v=IJxWoZJAD8k, and a number of articles and books on this topic

A Revolution in Manufacturing: The SMED System, S. Shingo, ISBN 978-0915299034

Product Design for Manufacturing and Assembly, G. Boothroyd, P. Dewhurst, ISBN 978-0824705848